=== Penn's ===
Grandest Cavern

The History, Legends and Description
of Penn's Cave
in Centre County, Pennsylvania

Compiled by **HENRY W. SHOEMAKER**
(Member of the Historical Societies of Berks and Snyder Counties)

THE STATUE OF LIBERTY

ENTRANCE TO PENN'S CAVE (Frontispiece)

Penn's Grandest Cavern

The History, Legends and Description of Penn's Cave in Centre County Pennsylvania

There is a cave
All overgrown with trailing, odorous plants,
Which curtain out the day with leaves and flowers,
And paved with veined emerald, and a fountain
Leaps in the midst with an awakening sound;
From its curved roof, the mountain's frozen tears,
Like snow or silver, or long, diamond spires,
Hang downward, raining forth a doubtful light;
And there is heard the ever-moving air,
Whispering without from tree to tree, and birds
And bees; and all around are mossy seats,
And the rough walls are clothed with long, soft grass.
                                    —Prometheus Unbound

(REVISED EDITION—ILLUSTRATED)

Published by the
ALTOONA TRIBUNE PRESS, Altoona, Pa.
════ 1916 ════

INDEX OF CHAPTERS.

I INTRODUCTION.

PENN'S CAVE needs more panegyrics and panegyrists. Beautiful natural curiosity that it is, it is hidden away among rolling hills and towering mountains, almost like "a flower to blush unseen." The writer of these lines, having visited many of the principal caves in the United States and in foreign countries, and comparing them with Penn's Cave, has come to the conclusion that something adequate *should* be written concerning the great Central Pennsylvania cavern. Though not having the spare time to go into the subject in detail, he has compiled the following chapters in the hopes of filling the want until the proper historian can take up the subect, using the contents of this book as a foundation for more sonu research and exposition. But this is sometimes a difficult task, as history loves to follow beaten paths. After much painstaking research and a world of care, the writer prepared the first complete history of the Pine Creek, or Fort Horn Declaration of Independence. It was a subject glossed over by most Pennsylvania historians, even by the immortal Meginness. A week or so ago in the "Romances of Pennsylvania History" series in a leading Philadelphia newspaper the old story was republished, just as it was given, fragmentary and imperfect in every old history. Either the compiler of the article disregarded a newer and more complete version, or did not see it, or else history is too dogmatic to leave its channels. In the case of Penn's Cave, its amplified story appears in these pages for the first time; it cuts out the channel, as it were. Consequently the writer feels an added responsibility, for here is a lack of the minuteness so characteristic of some other specimens of cave literature, notably Hovey's works. But in lieu of other treatises, these pages are presented to the public in the hope that they may answer a few of the questions being asked about the Cave, and to preserve the folk-lore clustered about it. To the writer these pages have a deep import, as Penn's Cave determined his course to collect and preserve, if possible, the dying legends and folk-tales of the Pennsylvania Mountains. Twenty-two years ago this month, as a little red-headed boy, he made the acquaintance of an aged Seneca Indian, Isaac Steele, who was visiting familiar scenes in the West Branch and Bald Eagle Valleys. The venerable man sat on the trunk of a felled Indian apple tree at the corner of the old Quiggle orchard, at McElhattan, (Clinton County), and recounted the "legend of Penn's Cave." For eleven years it tossed about in the writer's mind, until a time came when he could contain it no longer, so he wrote it down. It was first published in the "Centre Reporter," at Centre Hall, (Clinton County), and became the nucleus of other legends, which came out in book form in 1903, under the title of "Wild Life in Central Pennsylvania." Later editions of this book were published under the name of "Pennsylvania Mountain Stories," the last in 1911. And from that time on, when the writer had a little leisure, a little chance to travel, he has been collecting and "writing down" more Pennsylvania legends. Therefore, it is with more than the usual heartbeats that he is giving forth his latest brochure on "Penn's Grandest Cavern." The writer wishes to extend his hearty thanks to Mr. R. P. Campbell, one of the proprietors of the Cave, for valuable assistance rendered in the preparation of this book, and to Mr. S. W. Smith, editor of the "Centre Reporter," for furnishing some of the most interesting illustrations.

HENRY W. SHOEMAKER.

Altoona Tribune Office, Sept. 28, 1914.

II. DESCRIPTION.

THOUGH perhaps lacking in the exquisite stalactite formations of the Crystal Cave at Virgins-
ville, or the huge "dragon" stalagmite at
Dreibelbis Cave, or the "Red Panther's Funeral Pyre"
stalagmite in the Caves of Coburn, or the symmetry
of the bush-hammered walls of the Naginey Cave,
Penn's Cave excels all other Pennsylvania caverns by
the vastness of its dimensions, its water trip, its
diversity of formations. While other caves in the
Commonwealth rely on one feature of commanding
interest, Penn's Cave has first-class attractions by the
score. It contains so much that is of interest that it
always gives fresh and absorbing pleasure, even to
persons who have visited it dozens of times—like the
writer of this article. First of all, let it be said that
the entrance is the most imposing of any cave in the
United States—maybe in the world. The flight of
steps leading down to the vast limestone arch with the
depth of green water beneath it is something never to
be forgotten. The boat ride, a quarter of a mile or
more, each way, is finer by far than the Echo River,
the Styx or Lethe in Mammoth Cave in Kentucky,
or the Lake in Cahow Cave, or the boat ride in Smug-
glers' Cave in Bermuda, where Annette Kellerman
posed for the great moving picture play, "Neptune's
Daughter." The mysterious abruptness with which
Penn's Cave ends adds greatly to its charm. The

writer has crawled far into the labyrinths beyond the
ending of the "watery" part, been confused by the
multiplicity of passages, been lulled by the musical
echoes of countless subterranean waterfalls. The
Cave has three, possibly four or five, entrances. The
main entrance, already referred to, is like the doorway to the Labyrinth of the Minotaur in Auguste
Gendron's famous painting, the entrance from the old
orchard into the dry cave, and another which can be
noticed by the ray of light which filters into one of
the hidden chambers at the rear of "watery cave"
are   curiously   picturesque.   Other cork-screw or
spiral apertures are observable in the ceilings at certain parts of the cavern, but as they admit no light,
they cannot be definitely called "entrances." In some
places the water attains a depth of forty feet, and is
of a peculiar transparent greenish color. Trout and
other fish find their way into the Cave, but do not
multiply, as there is no food of importance inside,
yet the earlier explorers reported that it was fairly
alive with trout. Daniel Ott, of Selinsgrove, who died
recently, aged 96 years, has stated that at one time
shad were taken in Penn's Cave. In the coldest part
of winter screech owls take refuge in the cavern. Small
cray-fish, numerous insects, including white katydids,
rats, mice and bats, still inhabit it in considerable numbers. Unfortunately, the bats do not hibernate in as
great quantity as formerly. The noises, the acetylene
lights, the inquisitive tourists, have driven these shy
creatures to unknown hiding places, though their

desertion is not as complete as at the charming
Crystal Cave, in Berks County, where the proprietor
injudiciously installed a system of electric lighting.
At dusk in summer, beginning in May, numbers of
bats can be seen flitting about near the mouth of the
Cave and on the green before the hotel, chasing insects.
Thanks to their tireless efforts, there is an almost entire absence of mosquitoes in the vicinity of Penn's
Cave. No wonder Texas has put a perpetual closed season on our little satan-winged friends. It is said that,
their work done, the bats return to the Cave through
the small opening in the orchard, preferring it to the
larger, or main entrance. According to the stories told
by the first explorers the "Dry" Cave was formerly
much dryer than at present. In the old days, panthers,
red bears, lynxes, foxes, as well as smaller mammals,
made it a headquarters; the larger beasts fought for its
possession. Indians sometimes camped in the Dry
Cave in very cold winters. They would find it un-
comfortably wet now. The quality of the limestone
composing the walls of the Cave is very unusual. It
shades from whites to delicate greys, into rich pinks
and reds. It is the most gorgeously colored cavern in
the Eastern States. Italians might almost call it "the
American Capri." In some parts the delicacy of the
grey-green tones reminds one of the famous French
"art nouveau" introduced about the time of the Paris
Exposition in 1900. In other places the reds are rem-
iniscent of the richness of the best in Indian art. This
is best seen in the curious, natural mural painting

called "Indian Riding Pony," which is shown to visitors on the "return trip" in the Cave. It is a "sumac" red which holds the attention just as the primitive artists evidently sought, knowing that the impression must be given by only one color.† The stalactites are not as numerous as in many caves, notably Luray, the Endless Caverns near Newmarket, Va., or the Wyandot Cave on the Rothrock estate in Southern Indiana. Countless numbers were broken off during the dark days when there was no absolute rule in the Cave, when visitors did pretty much as they pleased. The early explorers spoke enthusiastically of the stalactites, so we must blame the generation of vandals if our Cave is exceeded in this respect by other American caverns. Some of the curious stalactite forms, like "The Lancaster County Tobacco Barn" and "The Lobster's Claw" are not to be excelled anywhere. But there are few transparent pendants, loveliest of all stalactites. The stalagmite forms are finer and more numerous than the stalactites, at least in their present-day condition. They represent a wide diversity of forms, some of them like the "Giant Pillars" being of impressive proportions, while others like the "Prairie Dogs" are quaint and amusing in the extreme. As a "freak" formation the "Ruffles, Scalloped" is well worth a visit. There are several places where the formations emit musical sounds upon being struck. One great charm of Penn's Cave is that the visitor never leaves disappointed, as is the case with many caves. The imagination, it

---

† See Chapter IX.

THE GARDEN OF THE GODS.

would seem, cannot picture anything quite like it. Many persons imagine Niagara Falls to be a grander sight than it seems to them on first glimpse, but on subsequent visits it appears to grow to the proportions of the preconceived mental image. Penn's Cave comes upon the eye very different from any prior conception; any subsequent visits make it seem lovelier, more weird, grander. Its situation in a picturesque region adds greatly to its attractiveness. Sixteen years ago, when the writer first visited the Cave, there was much original timber, white pine, white oak, hemlock, standing in the ravines adjacent to the property. Now, alas, much of this is gone, but there is still a quaint old-world, out-of-time atmosphere connected with the region. At night to lie on the hillside by the creek that runs from the cavern, as the writer has done, and watch the Brush Mountain above so immovable and vast, frowning like a tall sentinel upon the Cave property, while down in some sink a whippoorwill is improvising, or a fox barking on a distant "bench," is a rare treat to an impressionable soul which seeks the infinite. The Cave is best visited at dusk or after night-fall, if the full effect of the eerie surroundings is desired. The formations appear huger, the distances greater, the shadows more impenetrable, after darkness outside. Then to emerge again into the seemingly excessively warm air, into darkness, and hear a distant kildeer's mournful note or to see a bat flit mechanically over one's head, are experiences in keeping with one's bewitched mental attitude. The glory of the autumn coloring in the Cave woods,

or on the adjoining farm, is very wonderful. Fall is
the best time of the year to visit the great natural wonder. There are many hardwoods still standing, the
hickories in particular are radiantly yellow in Septem-
ber and October. Bluejays, newly arrived from the
north, cry out buoyantly. In May-time the orchard
and fields about the Cave are a mass of white and pink
sweet-scented blooms. Bird songs in the rising inflexion
are everywhere. The very earth smells sweet; world
hopes come into our breasts. But there is a tang in the
air in Autumn; it comes from the drying leaves, the
cracking nut burrs, the hardening earth, that gives us a
stronger grip on life. Nature is our friend in May-
time, we understand her mood, she seems helping us.
In the Autumn she appears to be drawing away, becoming more distant, forgetful of us. We reverence
her as more all-powerful; we feel more self-reliant.
Our imaginations for these reasons are soothed in
Spring, keenly awakened in the Fall. Apart from
Nature's grandeur, the wonders of the Cave formations hold us more "in chill October." How great a joy
if one could visit the Cave on Hallowe'en! Care should
be taken to have every wonderful formation pointed
out. They are well-named, not one can we afford to
miss. It is only after seeing them that a correct esti-
mate can be formed of Penn's Cave, its position in
relation to other caverns. We believe that the discern-
ing observer is bound to give it a very high place.
Although the lamented Rev. Horace C. Hovey omitted
mentioning it in his classic work, "Celebrated American

Caverns," published in 1882, it is not too late to record
Penn's Cave in the "underground hall of fame." Gradually its popularity is growing, its distinctive marvels
are showing out more boldly. Like a genius half understood, modest and retiring, it is coming to its own. The
Height of the roof in the highest part is 55 feet; the
water at its greatest depth 35 feet, at time of high
water. The temperature of the Cave, all the year
'round, is 50 degrees, and the Cave property is situated
at an altitude of 1,200 feet above sea level. The Cave
received its name because John Penn's Creek, formerly
named the Karoondinha, rises in it. Penn's Creek was
named after John Penn (1729-1795), grandson of the
founder of Pennsylvania, Captain James Potter having
given it this appellation in January, 1764. A legend of
one of John Penn's visits to Central Pennsylvania will
be found in the compiler's "More Pennsylvania Mountain Stories," Reading, 1912, in the chapter entitled
"Marsh Marigold." For the benefit of intending visitors, below is appended a list of the leading *named
formations,* 36 in all, but there are hundreds of others
which await their "great American identifiers," that
are full of strangeness, full of charm, tonics for the
imagination.

SEEN AS YOU ENTER CAVE.

1 The Eagle's Wings.
2 The Lobster's Claw.
3 Statue of Liberty.
4 A Bunch of Bananas.
5 Garden of Gods.
6 The Lace Curtain.
7 The Strait of Gibraltar.
8 Petrified Lion.
9 Coral Growths.
10 The Chimes.
11 Drop Curtain.
12 Prairie Dogs.
13 Snow Slides.
14 Pittsburg Snow Drift.
15 Niagara Falls (Canadian and American Sides).
16 Trout Colored Stalactite.
17 Turtle Shell.
18 Lancaster County Tobacco Barn.
19 Hindoo Idols.
20 Giant Pillars.

SEEN AS YOU RETURN.

1 Water Falls with Lighthouse Above.
2 The Ruffles, Scalloped.
3 North Pole Scene.
4 Indian Riding Pony.
5 Leopard Skins.
6 The Billiken.
8 Lebanon Bologna.
9 Boy Driving Cow Across Suspension Bridge.
10 Indian Woman Carying Papoose.
11 Egyptian Woman Carrying Jug of Water.
12 Dove Wing.
13 Angel Wings.
14 Silver Rock.
15 Shadow Statue of Libertty.
16 Elephant's Head.

AT THE PANTHER SKINS

III. HISTORY.

(From the Altoona Tribune.)

IT is not generally known that the namesakes, or perhaps distant relatives of America's greatest poet, Edgar Allan Poe, were the first white men to own Penn's Cave, in Centre County. These hardy frontiersmen, who fought the Indians in the mountains of Maryland and Pennsylvania, took up many tracts of land in the Pennsylvania mountains and became citizens of prominence. The original name was spelled Poh, but became altered like so many other of the old-time names, into its present form. The Penn's Cave farm, or tract of land, as it was known in the early days, was surveyed in pursuance of two warrants granted to James Poh or Poe and dated January 5 and November 3, 1773. A patent for these lands was issued by the Commonwealth of Pennsylvania to James Poe, dated April 9, 1789. James Poe only lived on the Penn's Cave farm a short time, spending most of his days at his homestead in the valley bearing his name in the southern part of Centre County. But he built a substantial log house near the large spring where the Karoondinha emerges from the cave, which was the first improvement in that part of the valley. James Poe at his death left the Cave farm to his daughter, Susanna M. Poe, and his will is duly recorded in the records of Franklin County, Pennsylvania, Centre County not yet having come into existence. The young

19

heiress became the wife of Samuel Vantries, and the
Penn's Cave farm took the name of the "Vantries
Place," by which name it was known for many years.
Samuel Vantries lived formerly near what is now
Linden Hall, as the family name is still well known
in that locality. Dr. James Vantries, of Bellefonte, is a
direct descendant of Samuel Vantries and James Poe.
There is no record that Edgar Allan Poe, during his
famous visit to Central Pennsylvania in 1838, ever paid
a visit to his namesakes at Penn's Cave. Prior to the
time of this trip he was residig in Philadelphia, and was
on the staff of the "Gentlemen's Magazine." He was in
need of money, being heavily in debt, and thought that
doubtless his wealthy namesakes in the mountains
would help him. He visited Poe Valley, and later
crossed the Seven Mountains to Milroy and Lewistown,
from which latter town he returned to Philadelphia.
He was much impressed with the large cave on the
Naginey farm near Milroy, and at the Mammoth
Spring on the Alexander farm, not far from Reedsville.
Samuel Vantries rented the Penn's Cave farm about
1855, as Jacob Harshbarger was living there at that
time. Mr. Harshbarger used to say that the first person to enter the cave was Rev. James Martin, a Presbyterian preacher, who died June 20, 1795, and is buried
on the Musser farm, near Penn Hall. Rev. Martin
was a native of Ireland, an honor graduate of Trinity
College, Dublin, and pastor of the earliest Presbyterian
congregation in Penn's Valley. The old gentleman, it
is said, caught a cold in the cave, from which he never

fully recovered. Previous to Rev. Martin's adventure,
Indians of various tribes had frequented it, as numerous sovenirs, like arrow-heads, pottery and beads, have
been taken out of it. Malachi Boyer, a young pioneer
from Lanaster County, was drowned in the cave about
1749. He had run away with Nita-nee, the daughter of
a powerful chief, Okocho, was captured and paid the
death penalty.* Beginning with 1845, and continuing
to 1860, people frequently went down into the dry cave
through the small entrance in the old orchard. No
guides accompanied the visitors, however, and on an
occasion a pair of saddle horses were found tied to the
orchard fence at dusk one evening, which bore the
marks of having been tied there for some time. A
search was made in the dry cave and a young man and
his sweetheart were found close by the water. Their
lights had failed them and they were afraid to move,
and they had lost all idea as to which way to turn to
get out. So they decided to wait and trust to the presence of the horses to bring relief.† About 1860 a young
Quaker named Isaac Paxton, who had resided in Ches-
ter County, became teacher at the public school in
Spring Mills. He was a nature lover and fond of
taking long tramps through the hills and valleys to
study the birds and flowers, trees and geological formations. Accompanied by his chum, Albert Woods, a
successful agriculturist residing at Spring Mills, he
walked to Penn's Cave and entered the dry cave. The

* See Chapter IV.
† See Chapter VI.

young men became convinced that they saw a light out in the direction of the water-course entrance. Previous to this time there was no knowledge of the water in the "dry" cave being the same stream that rises at the main entrance of the cave, nor that the two parts of the cave led into one another. Paxton and Woods came out of the dry cave, went down to the saw mill, which stood close to where the water emerges from the cavern, and from which water power it was run, and secured enough lumber to build a raft. They carried this lumber to the main, or present, entrance of the cave, nailed it together, and, with the aid of a pine torch and a long pole, traversed the water-course in Penn's Cave for the first time. They found that the water-way led into the dry cave, unearthed the skeletons of two huge panthers, and made other interesting discoveries.‡ Presbyterian preachers must have had a fondness for visiting caves, as a few days after Rev. J. E. Long, the Presbyterian pastor of the Valley, whose place of residence was at Hublersburg, in Nittany Valley, came over and, hearing of the adventure of Messrs. Paxton and Woods, persuaded them to repeat the trip, so that he might accompany them. So the three gentlemen returned to the cave, reconstructed the raft into a small boat and traversed the gloomy water-way. The news spread rapidly, and as the Fourth of July was approaching, a small picnic of members of "old-line" families was gotten up to spend the holiday at the cave and make use of the boat. Among those in the party

‡ See Chapter V.

were two aged ladies, Mrs. Margaret Foster and Miss Sarah Vanvalzah. Because of their venerable age the compliment was paid them of having the boat named for them, the "Sarah-Margaret." Among those in the merry party were Miss Mary Wilson, Miss Lizzie Cook, Miss Mary Duncan, Miss Mary Woods, Miss Ada Vanvalzah, Mrs. Robert Duncan, John Foster, John Wilson, Frank Vanvalzah, Harry Vanvalzah, Dr. John Woods, Robert Duncan and Miss Mary Buchanan . Miss Ada Vanvalzah later became the wife of Col. John A. Churchill, of St. Louis, a distinguished officer in the United States Army. Dr. John Woods practiced the profession of medicine at Boalsburg, Centre County, for many years. Miss Mary Woods, who is now living at Spring Mills, furnished the list of names of the happy party, most of whom are now enjoying their reward. Miss Ada Vanvalzah and Miss Mary Woods were the first ladies to enter the boat and go through the cave. All during the day one load would be rowed back as far as the dry cave in the rear of the cavern and left to explore the dry rooms while the boat returned for another load. For years following this picnic the country became so excited over the Civil War that little interest was taken in the cave until about 1870, when another picnic party visited the picturesque spot. This time the boat was hauled on a wagon from Beaver Dams, below Spring Mills, which in those days was a favorite spot for canoeists and boatmen generally. No signs were found of the old boat, the "Sarah-Margaret."

Previous to the last picnic, in 1868, Samuel Vantries sold the farm to George Long, who lived in the old farmhouse and used the water from the "spring," which in reality is the overflow from the cave. Mr. Long was a man of serious nature and objected strongly to pleasure-seekers entering the cave. Furthermore, he did not want people to contaminate what he now realized was his water supply. During his regime few people visited the cave. Upon his death, in 1884, the property passed into the hands of his two sons, Jesse and Samuel. These two young men had traveled extensively and realized the financial possibilities of the cave. It was worth much more than the farm, in their estimation. In their rambles they had visited the Mammoth Cave of Kentucky, which they declared was in no way superior to their own cavern. They built a larger boat and began charging admission to the cave. About 1885 they constructed the handsome building now known as the Penn's Cave Hotel. For a time they prospered, and hundreds of people visited their unique resort annually. In December, 1905, the farm was sold to John A. Herman, of Pleasant Gap, Centre County. In January, 1908, the farm and cave again changed hands and became the property of its present owners, Dr. H. C. and R. P. Campbell. Previous to this, for several years, owing to financial embarrassments, the Long brothers had abandoned the hotel and the place was deserted. The Campbell brothers, who are graduates of the Pennsylvania State College, and young men of education and foresight, improved the property extensively, making it one of the most unique resorts in Central Pennsylvania. To use the words of Mr. R. P. Campbell, who is the active manager of the hotel and cave, "Now has come the age of the automobile, and the cave again has become a place of interest to the tourists. The number of visitors has increased steadily each year since we bought the place, and we expect 1914 to be the banner year."† Penn's Cave is easy of access to residents of Altoona, especially those owning automobiles. Twenty years ago the Naginey Cave in Milroy was visited by Altoonans every Sunday during the summer months. On several occasions the Altoona Band waked the echoes of its dismal recesses. Now, since the automobile has come into use, Penn's Cave in Centre County can be reached as easily as was the Naginey Cave in the old days. The best way to reach Penn's Cave by automobile from Altoona is to follow the main road to Tyrone, thence to Wariror's Mark, Pennsylvania Furnace, Rock Springs and State College. From State College it is only a short distance to the cave over first-class roads. Those wishing to go by train can reach Spring Mills or Rising Springs Station, as it is called, on the Lewisburg and Tyrone Railroad, after changing cars at Bellefonte. Conveyances cannot always be obtained there, so that it would probably be better to go by train to State College from Bellefonte.

---

† This prophecy proved correct, as twice as many persons visited the cave as ever before. But all records were broken in July, 1916, when a veritable army of persons visited the cave.

There are several excellent liveries there, also automobiles which can be hired. The Campbell brothers, according to experienced travelers, know how to keep a hotel. They provide clean beds, baths and running water and other conveniences for their guests. Meals are furnished when ordered in advance. Their charges are moderate, especially when one considers the nature of the accommodations furnished. But what appeals mostly to tourists and automobile parties is the air of courtesy and politeness which pervades the place. Every one, from Mr. Campbell down, seems anxious to please, and the tired traveler will find nothing to ruffle his overstrained nerves. The scenery about the cave is magnificent; in fact, there is none finer in Central Pennsylvania. The Brush Mountain comes to an abrupt end east of the cave, while to the south looms the high peaks of the Seven Mountains' chain. Penn's Cave makes an ideal trip for Altoonans and gives them a chance to fully appreciate the matchless beauties of their native state. Filled with historic lore, it creates an impression never to be forgotten. Ex-Governor Curtin called it "Pennsylvania's greatest natural wonder."* Many distinguished persons have been entertained there, including parties of foreigners. All these traveled persons who ought to know have been loud in their praises of this grand spot. When the writer was last there, in the month of May, 1914, the new moon was shining brightly high above the Seven Mountains. It cast a ghostly light

* See Chapter VII.

over the old orchard which surrounds the commodious
hotel. Down in the deep gorge, where the green limestone water rushes from the cave, the whippoorwills
had begun their plaintive melodies. Never did nature
seem more beautiful than on this occasion. The entire
demesne seemed to radiate the spirit of long ago. A
mist was rising from the entrance to the cave, and as
night progressed it seemed to form into the figures of
the murdered Malachi Boyer and his sweetheart,
Nita-nee, surrounded by the hostile forms of the old
chief and his seven sons, who would not permit a
marriage between an Indian princess and a white ad-
venturer. Then these forms seemed to fade away,
and in their places came those of Rev. Martin, Rev.
Long, Paxton and Woods, the early explorers of the
cave. It was a night full of fancies and imaginings,
where one lives over his life in retrospect. It was a
place where one can find relief and rest from the
cares of the modern, complex life. If the fountain of
youth is in Pennsylvania, surely it must have flowed
out of the unsounded depths of Penn's Cave,† for all
who have been there have come away strengthened
and spiritually purified by its rare beauty and precious
flood of memories.

† See Chapter VII.

IV. THE LEGEND OF PENN'S CAVE.

(Related by Isaac Steele, an Aged Seneca Indian, in 1892.)

IN THE DAYS when the West Branch Valley was a trackless wilderness of defiant pines and submissive hemlocks twenty-five years before the first pioneer had attempted lodgment beyond Sunbury, a young Pennsylvania Frenchman, from Lancaster County, named Malachi Boyer, alone and unaided, pierced the jungle to a point where Bellefonte is now located. The history of his travels has never been written, partly because he had no white companion to observe them, and partly because he himself was unable to write. His very identity would now be forgotten were it not for the traditions of the Indians, with whose lives he became strangely entangled.

A short, stockily built fellow was Malachi Boyer, with unusually prominent black eyes and black hair that hung in ribbon-like strands over his broad, low forehead. Fearless, yet conciliatory, he escaped a thousand times from Indian cunning and treachery, and as the months went by and he penetrated further into the forests he numbered many redskins among his cherished friends.

Why he explored these boundless wilds he could not explain, for it was not in the interest of science, as he scarcely knew of such a thing as geography, and

28

it was not for trading, as he lived by the way. But on he forced his path, ever aloof from his own race, on the alert for the strange scenes that encompassed him day by day.

One beautiful month of April—there is no one who can tell the exact year—found Malachi Boyer camped on the shores of Spring Creek. Near the Mammoth Spring was an Indian camp, whose occupants maintained a quasi-intercourse with the pale-faced stranger. Sometimes old Chief O-ko cho would bring gifts of corn to Malachi, who in turn presented the chieftain with a hunting knife of truest steel. And in this way Malachi came to spend more and more of his time about the Indian camps, only keeping his distance at night and during religious ceremonies.

Old O-ko-cho's chief pride was centered in his seven stalwart sons, Hum-kin, Ho-ko-lin, Too-chin, Os-tin, Chaw-kee-bin, A-ha-kin, Ko-lo-pa-kin and his Diana-like daughter, Nita-nee. The seven brothers resolved themselves into a guard of honor for their sister, who had many suitors, among whom was the young chief E-Faw, from the adjoining sub-tribe of the A-caw-ko-tahs. But Nita-nee gently, though firmly, repulsed her numerous suitors, until such time as her father would give her in marriage to one worthy of her regal blood.

Thus ran the course of Indian life when Malachi Boyer made his bed of hemlock boughs by the gurgling waters of Spring Creek. And it was the first sight of her, washing a deer-skin in the stream, that led him to prolong his stay and ingratiate himself with her father's tribe.

Few were the words that passed between Malachi and Nita-nee, many the glances, and often did the handsome pair meet in the mossy ravines near the camp grounds. But this was all clandestine love, for friendly as Indian and white might be in social intercourse, never could a marriage be tolerated, until— there always is a turning point in romance—the black-haired wanderer and the beautiful Nita-nee resolved to spend their lives together, and one moonless night started for the more habitable East. All night long they threaded their silent way, climbing the mountain ridges, gliding through the velvet-soiled hemlock glades, and wading, hand in hand, the splashing, resolute torrents. When morning came they breakfasted on dried meat and huckleberries, and bathed their faces in a mineral spring. Until—there is always a turning point in romance—seven tall, stealthy forms, like animated mountain pines, stepped from the gloom and surrounded the eloping couple. Malachi drew a hunting knife, identical with the one he had given to Chief O-ko-cho, and, seizing Nita-nee around the waist, stabbed right and left at his would-be captors. The first stroke pierced Hum-kin's heart, and, uncomplainingly, he sank down dying. The six remaining brothers, although receiving stab wounds, caught Malachi in their combined grasp and disarmed him; then one brother held sobbing Nita-nee, while the others dragged fighting Malachi across the mountain.

That was the last the lovers saw of one another. Below the mountain lay a broad valley, from the center of which rose a circular hillock, and it was to this mound the savage brothers led their victim. As they approached, a yawning cavern met their eyes, filled with greenish limestone water. There is a ledge at the mouth of the cave, about six feet higher than the water, above which the arched roof rises thirty feet, and it was from here they shoved Malachi Boyer into the tide below. He sank for a moment, but when he rose to the surface, commenced to swim. He approached the ledge, but the brothers beat him back, so he turned and made for some dry land in the rear of the cavern. Two of the brothers ran from the entrance over the ridge to watch, where there is another small opening, but though Malachi tried his best, in the impenetrable darkness, he could not find this or any other avenue of escape. He swam back to the cave's mouth, but the merciless Indians were still on guard. He climbed up again and again, but was repulsed, and once more retired to the dry cave. Every day for a week he renewed his efforts to escape, but the brothers were never absent. Hunger became unbearable, his strength gave way, but he vowed he would not let the redskins see him die, so, forcing himself into one of the furthermost labyrinths, Malachi Boyer breathed his last.

Two days afterward the brothers entered the cave and discovered the body. They touched not the coins in his pockets, but weighted him with stones and dropped him into the deepest part of the greenish limestone water. And after these years those who have heard this legend declare that on the still summer nights an unaccountable echo rings through the cave, which sounds like "Nita-nee," "Nita-nee."

V. CAVE PANTHERS.

EVERYONE who has hunted in the "Seven Brothers," as the Seven Mountains are called in Central Pennsylvania, has heard of Daniel Karstetter, the famous Nimrod. *Though the greater part of a hundred years have passed since he was in his hey-day as a slayer of big game, his fame is undiminished. Anecdotes of his prowess are related in every hunting camp; by one and all he has been acclaimed the greatest hunter that the Seven Brothers ever produced. The great Nimrod, who lived to a very advanced age, was born on the banks of the Karoondinha, or Penn's Creek, at the Blue Rock, several miles back of the present town of Coburn. In addition to his hunting prowess he was interested in psychic experiences, and was as prone to discuss his adventures with supernatural agencies as his conflicts with the wild denizens of the forest. There was a particular ghost story which he loved dearly to relate. Accompanied by his younger brother Jacob he had been attending a dance one night across the mountains, in the environs of the town of Milroy, for like all the backwoods boys of his time, he was adept in the art of terpsichore. The long journey was made on horseback, the lads being mounted on stout Conestoga chargers. The homeward ride was commenced after

---

* The Seven Mountains comprise the Path Valley, Short, Bald, Thick Head, Sand, Shade andTussey Mountains.

33

midnight, the two brothers riding along the dark trail in single file. In the wide flat on the top of the "Big Mountain" Daniel fell into a doze. When he awoke, his mount having stumbled on a stone, Jacob was nowhere to be seen. Thinking that his brother had put his horse to a trot and gone on ahead, Daniel dismissed the matter of his absence from his mind. As he was riding down the steep slope of the mountain, he noticed a horseman waiting for him on the path. When they came abreast the other rider fell in beside him, skilfully guiding his horse so that it did not encounter the dense foliage which lined the narrow way. Daniel supposed the party to be his brother, although the unknown kept his lynx-skin collar turned up, and his felt cap was pulled down level with his eyes. It was pitchy dark, so to make sure Daniel called out, "Is that you, Jacob?" His companion did not reply, so the young man repeated his query in still louder tones, but all he heard was the crunching of the horses' hoof on the pebbly road.

Daniel Karstetter, master-slayer of panthers, red bears and lynxes, was no coward, though on this occasion he felt uneasy. Yet he disliked picking a quarrel with the silent man at his side, who clearly was not his brother, and he feared to put his horse to a gallop on the steep, uneven roadway. The trip home never before seemed of such interminable length. For the greater part of the distance Daniel made no attempt to converse with his unsociable comrade. Finally, he heaved a sigh of relief when he saw a light gleaming

THE GARDEN OF THE GODS

in the horse stable at the home farm. When he reached the barnyard gate he dismounted to let down the bars, while the stranger apparently vanished in the gloom. Daniel led his mount to the horse stable, where he found his brother Jacob sitting by the old tin lantern, fast asleep. He awakened him and asked him when he had gotten home. Jacob stated that his horse had been feeling good, so he let him canter all the way. He had been sleeping, but judged that he had been home at least half an hour. He had met no horseman on the road. Daniel was convinced that his companion had been a ghost, or, as they are called in the "Seven Brothers," a *gshpook*. But he made no further comment that night. A year afterwards, in coming back alone from a dance in Stone Valley, he was again joined by the silent horseman, who followed him to his barnyard gate. He gave up going to dances on that account. At least once a year, or as long as he was able to go out at night, he met the ghostly rider. Sometimes, when tramping along on foot after a hunt, or, in later years, coming back from market in his Jenny Lind, he would find the silent horseman at his side. After the first experience he never attempted to speak to the nightrider, but he became convinced that it meant him no harm. As his prowess as a hunter became recognized he had many jealous rivals among the less successful Nimrods. In those old days threats of all kinds were freely made. He heard on several occasions that certain hunters were setting out to "fix" him. But a man who could

wrestle with panthers and bears knew no such thing
as fear. One night, while tramping along in Green's
Valley, he was startled by someone in the path ahead
of him shouting out in Pennsylvania German, "Hands
up!" He was on the point of dropping his rifle, when
he heard the rattle of hoofbeats back of him. The
silent horseman in an instant was by his side, the
dark horse pawing the earth with his giant hoofs.
There was a crackling of brush in the path ahead, and
no more threats of *hend uff*. The ghostly rider followed Daniel to his barnyard gate, but was gone before he could utter a word of thanks. As the result
of this adventure he became imbued with the idea
that he possessed a charmed life. It gave him added
courage in his many encounters with panthers, the
fierce red bears and lynxes.

Apart from his love of hunting the more dangerous
animals, Daniel enjoyed the sport of deer-shooting. He
maintained several licks, one of them in a patch of low
ground near the entrance to the "dry" part of Penn's
Cave. At this spot he constructed a blind, or platform,
between two ancient tupelo trees, about twenty feet
from the ground, and many were the huge white-
faced stags which fell to his unerring bullets during
the rutting season. One cold night, according to an
anecdote frequently related by one of his descendants,
while perched in his eyrie overlooking the natural
clearing which constituted the lick, and in sight of a
path frqeuented by the fiercer beasts, which led to the
opening of the "dry" cave, he saw, about midnight,

a huge pantheress, followed by a large male of the
same species, come out into the open. "The pantheress
strolled from the path," so the story went, "and came
and laid herself down at the roots of the tupelo trees,
while the panther remained in the path and seemed
to be listening to some noise as yet inaudible to the
hunter. Daniel soon heard a distant roaring; it seemed
to come from the very summit of the Brush Mountain, and immediately the pantheress answered it.
Then the panther, on the path, his jealousies aroused,
commenced to roar with a voice so loud that the
frightened hunter almost let go his trusty rifle and
held tighter to the railing of his blind, lest he might
tumble to the earth. As the voice of the animal that
had heard in the distance gradually approached,
the pantheress welcomed him with renewed roarings,
and the panther, restless, went and came from the path
to his flirtatious flame, as though he wished her to keep
silence, and from the pantheress to the path, as though
to say, 'Let him come if he dares; he will find his
match.' In about an hour a gigantic panther stepped
out of the forest and stood in the full moonlight on
the other side of the cleared place. The pantheress,
eyeing him with admiration, raised herself to go to
him, but the panther, divining her intent, rushed before her and marched right at his adversary. With
measured step and slow, they approached to within a
dozen paces of each other. Their smooth, round heads
high in the air, their bulging yellow eyes gleaming
their long, tufted tails slowly sweeping down the

brittle asters that grew about them. They crouched
to the earth—a moment's pause—and then they
bounded with a hellish scream high in the air and
rolled on the ground, locked in their last embrace.
The battle was long and fearful to the amazed and
spellbound witness of this midnight duel. Even if he
had so wished, he could not have taken steady enough
aim to fire. But he preferred to watch the combat,
while the moonlight lasted. The bones of the two
combatants cracked under their powerful jaws, their
talons strewed the frosty ground with entrails, and
painted it red with blood, and their outcries, now
gutteral, now sharp and loud, told their rage and
agony. At the beginning of the contest the pantheress
crouched herself on her belly, with her eyes fixed
upon the gladiators, and all the while the battle raged,
manifested by the slow, cat-like motion of her tail
the pleasure she felt at the spectacle. When the
scene closed, and all was quiet and silent and death-
like on the lick and the moon had commenced to wane,
she cautiously approached the battle-ground, and,
sniffing the lifeless bodies of her two lovers, walked
leisurely to a nearby oak, where she stood on her hind
feet, sharpening her fore claws on the bark. She
glared up ferociously at the hunter in the blind, as if
she meant to vent her anger by climbing after him. In
the moonlight her golden eyes appearing so terrifying
that Daniel dropped his rifle and it fell to the earth
with a sickening thud. As he reached after it the
flimsy railing gave way and he fell, literally into the

arms of the pantheress. Just at that moment the rumble of horses' hoofs, like thunder on some distant mountain, was heard. Just as the panther was about to rend the helpless Nimrod to bits, the unknown rider came into view. Scowling at the intruder, mounted on his huge black horse, the brute abandoned her prey and ambled off in the direction of the dry cave. Daniel seized his firearm and sent a bullet after her retreating form, but it apparently went wild of its mark. Mean while, before he had time to express his gratitude to the strange deliverer he had vanished. Daniel was dumbfounded. As soon as he had recovered from the blood-curdling episodes, he built a small fire near the mammoth carcasses, where he warmed his much benumbed hands. Then he examined the dead panthers, but found that their hides were too badly torn to warrant skinning. Disgusted at not getting his deer, and being even cheated out of the panther pelts, he dragged the ghastly remains of the erstwhile kings of the forest by their tails to the edge of the entrance to the dry cave. There he cut off the long ears in order to collect the bounty, and then shoved the carcasses into the aperture. They fell with sickening thuds into the chamber beneath, to the evident horror of the pantheress, which uttered a couple of piercing screams as the horrid remains of the recent battle royal landed in her vicinity. Then Jacob shouldered his rifle and started out in search of small game for his breakfast. That night he went to another of his licks on Elk Creek, where he killed four superb stags," so the story concludes. But to his dying day he always placed the battle of the panthers first of all his hunting adventures. And his faith in the unknown horseman as his deliverer and good genius became the absorbing, all-pervading influence of his life.

VI. THE LOST LOVERS.

IT WAS long past dark when Mifflin Sargeant, of
the Snow Shoe Land Company, came within sight
of the welcoming lights of Stover's. For eighteen
miles, through the foothills of the Narrows, he had
not seen a sign of human habitation, except one deserted hunter's cabin. There was an air of cheerful-
ness and life about the building he had arrived at.
Several doors opened simultaneously at the signal of
his approach, given by a faithful watchdog, throwing
the rich glow of the fat lamps and tallow candles across
the road. The structure, which was very long and
two stories high, housed under its accommodating
roof a tavern, a boarding house, a farmstead, a lumber
camp, a general store and a post office. It was the last
outpost of civilization in the east end of Brush Valley;
beyond were mountains and wilderness almost to
Youngmanstown. Tom Tunis had not yet erected the
substantial structure in the midst of the forest later
known as "The Forest house." A dark-complexion-
ed lad, who later proved to be the son of the land-
lord, took the horse by the bridle, assisting the young
stranger to dismount. He also helped him to un-
strap his saddle-bags, carrying them into the house.
Sargeant noticed, as he passed across the porch, that
the walls were closely hung with stags' horns, which
showed the prevalence of these noble animals in the
neighborhood. Old Daddy and Mammy Stover, who

ran the quaint caravansery, quickly made the visitor
feel at home. It was after the regular supper-time,
but a fresh repast was cheerfully prepared in the huge
stone chimney. The young man explained to his hosts
that he had ridden that day from New Berlin; he had
come from Philadelphia to Harrisburg by train, to
Liverpool by packet boat, at which last named place
his horse had been sent on to meet him. He added
that he was on his way into Centre County, where
he had recently purchased an interest in the Snow
Shoe development. After supper he strolled along
the porch to the far end, to the post office, thinking
he would send a letter home. A mail had been brought
in from Rebersburg during the afternoon, consequently the post office, and not the tavern stand, was
the attraction of the crowd this night. The narrow
room was poorly lighted by fat lamps, which cast
great, fitful shadows, making grotesques out of the
oddly-costumed, bearded wolf hunters present, who
were the principal inhabitants of the surrounding
ridges. A few women, hooded and shawled, were
noticeable in the throng. In a far corner, leaning
against the water bench, was young Reuben Stover,
the hostler, tuning up his wheezy fiddle. As many
persons as possible hung over the rude counter, across
which the mail was being delivered, and where many
letters were written in reply. Above this counter
were suspended three fat lamps, attached to grooved
poles, which, by cleverly-devised pulleys, could be
lifted to any height desired. The young Philadelphian

THE CHIMES

VIEW ALONG THE KAROONDINHA

edged his way through the good-humored concourse
to ask permission to use the ink; he had brought his
favorite quill pen and the paper with him. This
brought him face to face, across the counter, with
the postmistress. He had not been able to see her before, as her trim little figure had been wholly ob-
scured by the ponderous forms that lined the counter.
Instantly he was charmed by her appearance—it was
unusual—by her look of neatness and alertness. Their
eyes met—it was almost with a smile of mutual
recognition. When he asked her if he could borrow
the ink, which was kept in a large earthen pot of
famous Sugar Valley make, she smiled on him again,
and he absorbed the charm of her personality anew.
Though she was below the middle height, her figure
was so lithe and erect that it fully compensated for
the lack of inches. She wore a blue homespun dress,
with a neat checked apron over it, the material for
which constituted a luxury, and must have come all
the way from Youngmanstown or Sunbury. Her profuse masses of soft, wavy, light-brown hair, on which
the hanging lamps above brought out a glint of gold,
was worn low on her head. Her deepset eyes were
a transparent blue, her features, well developed, and
when she turned her face in profile, the high arch of
the nose showed at once mental stability and energy.
Her complexion was fair; there seemed to be always
that kindly smile playing about the eyes and lips. When
she pushed the heavy inkwell towards him he noticed
that her hands were very white, the fingers tapering;

they were the hands of innate refinement. Almost
imperceptibly the young man found himself in conver-
sation with the little postmistress. Doubtless she was
interested to meet an attractive stranger, one from
such a distant city as Philadelphia. While they talked,
the letter was gradually written, sealed, weighed and
paid for; it was before the days of postage stamps,
and the postmistress politely waited on her customers.
He had told her his name—Mifflin Sargeant—and she
had given him hers—Caroline Hager—and that she
was eighteen years of age. He had told her about his
prospective trip into the wilds of Centre County, of
the fierce beasts which he had heard still abounded
there. The girl informed him that he would not have
to go farther west to meet wild animals; that wolf
hides by the dozens were brought to Stover's each
winter, where they were traded in; that old Stover, a
justice of the peace, attested to the bounty warrants—
in fact, the wolves howled from the hill across the
road on cold nights when the dogs were particularly
restless. Her father was a wolf hunter, and would
never allow her to go home alone; consequently, when
he could not accompany her, she remained in the
dwelling which housed the post office. Panthers, too,
were occasionally met with in the locality, also huge
red bears and the somewhat smaller black ones. If
he was going west, she continued in her pretty way,
he must not fail to visit the great limestone cave near
where the Brush Mountain ended. She had a sister
married and living not far from it, from whom she

had heard wonderful tales, though she had never been there herself. It was a cave so vast it had not as yet been fully explored; one could travel for miles in it in a boat; John Penn's Creek had its source in it; Indians had formerly lived in the dry parts, and wild beasts. Then she lowered her voice to say that it was now haunted by the Indians' spirits. And so they talked until a very late hour, the crowd in the post office melting away, until Jared Hager, the girl's father, in his wolfskin coat, appeared to escort her home, to the cabin beyond the waterfall near the trail to Hope Valley. She was to have a holiday until the next afternoon. The wolf hunter was a courageous-looking man, much darker than his daughter, with a heavy beard and bushy eyebrows. He spoke pleasantly with the young stranger, and then they all said good night. "Don't forget to visit the great cavern," Caroline called to the youth. "I surely will," he answered, "and stop here on my way east to tell you all about it." "That's good; we want to see you again," said the girl, as she disappeared into the gloomy shadows which the shaggy white pines cast across the road. Young Stover was playing "Green Grows the Rushes" on his fiddle in the tap-room and Sargeant sat there listening to him, dreaming and musing all the while, his consciousness singularly alert, until the closing hour came. That night, in the old stained four-poster, in his tiny, cold room, he slept not at all. "Yet he feared to dream." Though his thoughts carried him all over the world, the little postmistress was uppermost in every fancy. Among other things, he wished that he had asked her to ride with him to tne cave. They could have visited the subterranean marvels together. He got out of bed and managed to light the fat lamp. By its sputtering gleams he wrote her a letter, which came to an abrupt end as the small supply of ink which he carried with him was exhausted. But as he repented of the intense sentences penned to a person who knew him so slightly, he arose before morning and tore it to bits. There was a white frost on the buildings and ground when he came downstairs. The Autumn air was cold, the atmosphere was a hazy, melancholy grey. There seemed to be a cessation of all the living forces of nature, as if waiting for the summons of winter. From the chimney of the old inn came the pungent odor of burning pine wood. With a strange sadness he saddled his horse and resumed his ride towards the west. He thought constantly of Caroline—so much so that after he had traveled ten miles he wanted to turn back; he felt miserable without her. If only she were riding beside him, the two bound for Penn's Valley Cave, he could be supremely happy. Without her, he did not care to visit the cavern, or anything else; so at Madisonburg he crossed the northern mountains, leaving the southerly valleys behind. He rode up Nittany Valley to Bellefonte, where he met the agent of the Snow Shoe Company. With this gentleman he visisted the vast tract being opened up to lumbering, mining and colonization. But his thoughts were elsewhere; they were across the mountains with the little postmistress of Stover's. Satisfied that his investment would prove remunerative, he left the development company's cozy lodge-house, and, with heart growing lighter with each mile, started for the East. It was wonderful how differently—how vastly more beautiful the country seemed on this return journey. He fully appreciated the wistful loveliness of the fast-fading Autumn foliage, the crispness of the air, the beauty of each stray tuft of asters, the last survivors of the wild flowers along the trail. The world was full of joy, everything was in harmony. Again it was after nightfall when he reined his horse in front of Stover's long, rambling house. This time two doors opened simultaneously, sending forth golden lights and shadows. One was from the tap-room, where the hostler emerged; the other from the post office bringing little Caroline. There was no mail that night, consequently the office was practically deserted; she had time to come out and greet her much-admired friend. And let it be said that ever since she had seen him, her heart was aflame with the image of Mifflin Sargeant. She was canny enough to appreciate such a man, besides he was a good-looking youth, though perhaps of a less robust type than those most admired in the Red Hills. After cordial greetings the young man had his supper, after which he repaired to the post office. By that time the last straggler was gone; he had a blissful evening with his fair Caroline. She anticipated his coming, being somewhat of a psychic, and had arranged to spend the night with the

Stovers. There was no hurry to retire; when they
went out on the porch preparatory to locking up, the
hunter's moon was sinking behind the western knobs,
which rose like the pyramids of Egypt against the sky
line. Sargeant lingered around the old house for three
days; when he departed it was with extreme reluct-
ance. Seeing Caroline again in the future appeared
like something too good to be true, so downhearted
was he at the parting. But he had arranged to come
back the following Autumn, bringing an extra horse
with him, and the two would ride to the wonderful
cavern in Penn's Valley and explore to the ends its
stygian depths. Meanwhile they would make most
of their separation through a steady correspondence.
Despite glances, pressure of hands, chance caresses,
and evident happiness in one another's society, not a
word of love had passed between the pair. That was
why the pain of parting was so intense. If Caroline
could have remembered one loving phrase, then she
would have felt that she had something tangible on
which to hang her hopes. If the young Philadelphian
had unburdened his heart by telling her tnat he loved
her, and her alone, and heard her words of affirma
tions, the world out into which he was riding would
have seemed less a blank. But underneath his love,
burning like a hot branding iron, was his consciousness
of class, his fear of the consequences if he took to the
great city a bride from another *sphere*. As an only
son, he could not picture himself deserting his widowed
mother and sisters and living at Snow Shoe; there he

was sure that Caroline would be happy. Neither could
he see permanent peace of mind if he married her and
brought her into his exclusive circles in the Quaker
City. As he was an honorable young man, and his
love was real, making her truly and always happy was
the solitary consideration. These thoughts marred
the parting; they blistered and ravaged his spirit on
the whole dreary way back to Liverpool. There his
colored servant, an antic darkey, was waiting to ride
the horse to Philadelphia. The young man boarded
the packet, riding on it to Harrisburg, where he took
the steam train for home. In one way he was happier
than ever before in his life, for he had found love;
in another he was the most dejected of men, for his
beloved might never be his own. He seemed gayer
and stronger to his family; evidently the trip into the
wilderness had done him good. He had begun his
letter-writing to Caroline promptly. It was his great
solace in his heart perplexity. She wrote a very good
letter, very tender and sympathetic, the handwriting
was clear, almost masculine, denoting the bravery of
her spirit. During the winter he was called upon
through his sisters to mingle much with the society of
the city. He met many beautiful and attractive young
women, but for him the die of love had been cast. He
was Caroline's irretrievably. Absence made his love
firmer, yet the solution of it all the more enigmatical.
The time passed on apace. Another Autumn set in,
but on account of important business matters it was
not until December that Sargeant departed for the

wilds of Central Pennsylvania. But he could spend Christmas with his love. This time he sent two horses ahead to Liverpool. When he reached the queer old river town he dropped into an old saddlery shop, where the canal-boat drivers had their harness mended, and purchased a neat side saddle all studded with brass headed nails. This he tied on behind his servant's saddle. The two horsemen started up the Mahantango, crossing the Shade Mountain to Swine-fordstown, thence to Hartley Hall and the Narrows, a slightly shorter route to Stover's. On his previous trip he had ridden along the river to Selin's Grove, across Chestnut Ridge to New Berlin, over Shamokin Ridge to Youngmanstown, and from there to the Narrows; he was in no hurry; no dearly loved girl was waiting for him in those days. Caroline, looking prettier than ever—she was a trifle plumper and redder cheeked—was at the post office steps to greet him. Despite his avoidance of words of love, she was certain of his inmost feelings, and opined that somehow the ultimate result would be well. Sargeant had ar-ranged to arrive on a Saturday evening, so that they could begin their ride to the cave that night after the post office closed, and be there bright and early Sun-day morning. For this reason he had traveled by very easy stages from Hartley Hall, that the horses might be fresh for their added journey. Sargeant's devoted Negro factotum was taken somewhat aback when he saw how attentive the young man was to the girl, and marveled at the mountain maid's rare beauty.

THE GHOST ROOM.

Upon instructions from his master, he set about to
changing the saddles, placing the brand new lady's
saddle on the horse he had been riding. It was not
long until the tiny post office was closed for the night,
and Caroline emerged, wearing a many-caped red
riding coat, the hood of which she threw over her
head to keep the wavy, chestnut hair in place. She
climbed into the saddle gracefully—she seemed a natural horsewoman—and soon the loving pair were can-
tering up the road towards Wolf's Store, Rebersburg
and the cave. It was not quite daybreak when they
passed the home of old Jacob Harshbarger, the tenant
of the "cave farm;" a Creeley rooster was crowing
lustily in the barnyard, the unmilked cattle of the
ancient black breed shook their heads lazily; no one
was up. The young couple had planned to visit the
cave, breakfast and spend the day with Caroline's sister, who lived not far away at Centre Hill, and ride
leisurely back to Stover's in the late afternoon. It
had been a very cold all-night ride, but they had been
so happy that it had seemed brief and free from all disagreeable physical sensations. In those days there was
no boat in the cave, and no guides; consequently all
intending visitors had to bring their own torches. This
Caroline had seen to, and in her leisure moments for
weeks before her lover's coming, had been arranging
a supply of rich-pine lights that would see them safely
through the gloomy labyrinths. They fed their horses
and then tied them to the fence of the orchard which
surrounded the entrance to the "dry" cave, and which

had been recently set out. Several big original white
pines grew along the road, and would give the horses
shelter in case it turned out to be a windy day. The
young couple strolled through the orchard, and down
the steep path to the mouth of the "watery" cave,
where they gazed for some minutes at the expanse of
greenish water, the high span of the arched roof, the
general impressiveness of the scene. so like the stage-
setting of some elfin drama. They sat on the dead
grass, near this entrance, eating a light breakfast
with relish. Then they wended their way up the hill
to the circular "hole in the ground" which formed the
doorway to the dry cave. The torches were carefully
lit, the supply of fresh ones was tied in a bundle about
Sargeant's waist. The burning pine gave forth an
aromatic odor and a mellow light. They descended
through the narrow opening, the young man going
ahead and helping his sweetheart after him. Down
the spiral passageway they went, until at length they
came into a large chamber. Here the torches cast un-
earthly shadows, bats flitted about; some small animal
ran past them into an aperture at a far corner. Sargeant declared that he believed the elusive creature a
fox, and he followed in the direction in which it had
gone. When he came to this opening he peered
through it, finding that it led to an inner chamber of
impressive proportions. He went back, taking Caroline by the hand, and led her to the narrow chamber,
into which they both entered. Once in the interior room
they were amazed by its size, the height of its roof,

the beauty of the stalactite formations. They sat down
on a fallen stalagmite, holding aloft their torches, absorbed by the beauty of the scene. In the midst of
their musing a sudden gust of wind blew out their
lights. They were in utter darkness. The young
lover bade his sweetheart be unafraid, while he
reached his hand in his pocket for the matches. They
were primitive affairs, the few he had, and he could
not make them light. He had not counted on the use
of the matches, as he thought one torch could be lit
from another; consequently had brought so few with
him. Finally he lit a match, but the dampness extin-
guished it before he could ignite his torch. When the
last match failed, it seemed as if the couple were in a
serious predicament. They first shouted at the top of
their voices, but only empty echoes answered them.
They fumbled about in the chamber, stumbling over
rocks and stalagmites, their eyes refusing to become
accustomed to the profound blackness. Try as they
would, they could not locate the passage that led from
the room they were in to the outer apartment. Caroline, little heroine that she was, made no complaint. If
she had any secret fears her lover effectually
quenched them by telling her that the presence of the
two saddle horses tied to the orchard fence would
acquaint the Harshbarger family of their presence in
the cave. "Surely," he went on, "we will be rescued
in a few hours. There's bound to be some member of
the household or some hunter see those horses." But
the hours passed, and with them came no intimations

of rescue. But the two "prisoners" loved one another, time was as nothing to them. In the outer world, both thought, but neither made bold to say, that they might have to separate—in the cave they were one in purpose, one in love. How gloriously happy they were. But they did get a trifle hungry, but that was appeased at first by the remnants of the breakfast provisions, which they luckily still had in a little bundle. When sufficient time had elapsed for night to set in, they fell asleep, and in each other's arms. Caroline's last consciuos moment was to feel her lover's kisses. When they awoke, many hours afterwards, they were hungrier than ever, and thirsty. Sargeant fumbled about, locating a small pool of water, where the two quenched their thirst. But still they were happy, come what may. They would be rescued, that was certain, unless the horses had broken loose and run away, but there was small chance of that. They had been securely tied. It was strange that no one had seen the steeds in so long a time, with the farmhouse less than a quarter of a mile away—but it was at the foot of the hill. Hunger grew apace with every hour After a while drinking water would not sate it. It throbbed and ached, it became a dull pain, that only love could triumph over. Again enough hours elapsed to bring sleep, but it was harder to find repose, though Sargeant's kisses were marvelous recompense. Caroline never whimpered from lack of food. To be with her lover was all she asked. She had prayed for over a year to be with him again. She

would be glad to die at his side, even of starvation.
The young man was content; hunger was less a pain
to him than had been the past fourteen months' separation. Again came what they supposed to be morning. They knew that there must be some way out
near at hand, as the air was so pure. They shouted,
but the dull echoes were their only reward. Strangely
enough, they had never felt another cold gust like
the one which had blown out their torches. Could the
shade of one of the old-time Indians who had fought
for possession of the cave been perpetrator of the
trick, suggested lovely little Caroline. If so, she
thought to herself; he had helped her, not harmed her,
for could there be in the world a sensation half so
sweet as sinking to rest in her handsome lover's arms?
Meanwhile the world outside the cavern had been going its way. Shortly after the young equestrians
passed the Harshbarger dwelling, all the family had
come out, and, after attending to their farm duties,
driven off to the Seven Mountains, where the sons of
the family maintained a hunting camp on the Karoondinha on the other side of High Valley. The boys
had killed an elk, consequently the guests remained
longer than expected, to partake of a grand Christmas
feast. They tarried at the camp all of that day, all
of the next; it was not until early on the morning of
the third day that they started back to the Penn's
Creek farm. They had arranged with a neighbor's
boy, Mosey Shell. who lived further along the creek
below the farmhouse, to do the feeding in their ab-

sence; it was winter, there was no need to hurry home.
When they got home they found Mosey in the act of
watering two very dejected and dirty looking horses
with saddles on their backs. "Where did *they* come
from " shouted the big freight-wagon load in unison.
"I found them tied to the fence up at the orchard.
By the way they act I'd think they hadn't been watered
or fed for several days," replied the boy. "You
dummy!" said old Harshbarger, in Dutch. "Some-
body's in that cave, and got lost, and can't get out."
He jumped out of the heavy wagon and ran to a corner of the corncrib, where he kept a stock of torches.
Then he hurried up the steep hill towards the entrance
to the dry cave. The big man was panting when he
reached the opening, where he paused a moment to
kindle a torch. Then he lowered himself into the pit,
shouting at the top of his voice, "Hello! Hello'
Hello!" It was not until he had gotten into the first
chamber that the captives in the inner room could hear
him. Sargeant had been sitting with his back
propped against the cavern wall, while Caroline, very
pale and white-lipped, was lying across his knees,
gazing up into the darkness, imagining that she could
see his face. When they heard the cheery shouts of
their deliverer, they did not instantly attempt to
scramble to their feet. Instead the young lover bent
over; his lips touched Caroline's, who instinctively had
raised her face to meet his. As his lips touched hers,
he whispered, "I love you, my darling, with all my
heart. We will be married when we get out of here."

Caroline had time to say, "You are my only love," before their lips came together. They were in that position when the flare of Farmer Harshbarger's torch lit
up their hiding place. Pretty soon they were on their
feet, and, with their rescuer, figuring out just how
long they had been in their prison—their prison of
love. They had gone into the cave on the morning of
December 24th; it was now the morning of the 27th;
in fact, almost noon. Christmas had come and gone.
Caroline still had enough strength in reserve to enable
her to climb up the tortuous passage, though her lover
did help her some, as all lovers should. The farmer's
wife had some coffee and buckwheat cakes ready when
they arrived at the manse, which the erstwhile captives
of Penn's Cave sat down to enjoy. As they were eating, another of Harshbarger's sons rode up on horseback. He had been to the postoffice at Earlysburg. He
handed Sargeant a tiny, badly typed newspaper pub-
lishehed in Millheim. Across the front page, in letters
larger than usual, were the words, "Mexico Declares
War With the United States." Sargeant scanned the
headline intently, then laid the paper on the table. "Our
country has been drawn into a war with Mexico," he
said, with a voice trembling with emotion. "I had hoped
it might be avoided. I am First Lieutenant of the
Greys; I fear I'll have to go." Caroline lost the color
which had come back to her pretty cheeks since emerg-
ing from the underground dungeon. She reached over,
grasping her lover's now clammy hand. Then, noticing that no one was listening, she said faintly: "It is

terrible to have you leave me now; but won't you marry me before you go? I do love you." "Certainly I will," replied Sargeant, with enthusiasm. "I will have more to fight for, with you at home bearing my name."

AN INTERIOR VIEW OF DRY CAVE

VII. GOV. CURTIN'S VISIT.

CAPTAIN JOHN Q. DYCE, one of the pioneer
Democratic leaders of Clinton County, who died
in 1904, was fond of telling about Governor
Andrew G. Curtin's visit to Penn's Cave, and' the
great statesman's opinion of the cavern. It appeared
that during the Philadelphia Centennial in 1876 among
hosts of other celebrated foreigners who visited the
exposition were three Russians of note, Field Marshal
von Fersen, Count Hickoff, and Baron de Toplitz-
Harberstain. They were accompanied by their secretaries and retinues of servants. The heat of the city
was intolerable, it was in the month of August, and,
tiring of the marvels of the exposition, they sought to
visit the interior of the state in search of cooler
weather. One of the party recalled the fact that a
few years previously Andrew G. Curtin, who lived
somewhere in Central Pennsylvania, had been in Russia as United States Minister. The Russians admired' the gallant "War Governor," who had made a
most efficient envoy, so nothing would satisfy them
but to seek him out and pay their respects. And thus
it came to pass that one night, when the Bald Eagle
Valley train pulled into Bellefonte, it deposited on the
platform, to the wonder of the collected natives, three
Russian grandees, nine lesser individuals, and a pile
of luggage mountain high. It so happened that ex-
Governor Curtin was at home alone, the rest of his

family being at Saratoga. The ticket agent informed
the great statesman that some foreigners, who spoke
very little English, were waiting for him at the sta
tion. Hurrying to the depot as fast as he could travel,
he recognized his intending guests, who embraced him
in turn. They accepted the proffered invitation to
spend the night at the War Governor's mansion, and
soon the entire party was riding up the hill in a hotel
bus, commandeered for the purpose. Once in the
commodious mansion, the Russians felt perfectly at
home. First of all, they salaamed many times before
Brookman's magnificent oil portrait of the Czar
Nicholas II., which the "Little Father" had graciously
presented to Curtin before his departure from St.
Petersburg, and which hung in the War Governor's
library. The visitors were much impressed by the dry,
cool, pine-laden air, which reminded them, they said,
so much of Russia. These remarks made the tactful
Curtin decide that the best form of entertainment
would be a drive into the surrounding country. With
his truly matchless memory he recollected that Count
Hickoff was a man of some scientific attainments, had
been one of the party to unearth the skeleton of a
mammoth in Siberia, the tusks of which had been sent
to the Stuttgart museum, measured on the outside
curve twelve feet ten and one-half inches, and had a
greatest circumference of thirty-one and one-half
inches. Doubtless the noblemen would enjoy an ex-
cursion to the Penn's Cave, situated within a delightful driving distance of Bellefonte. Captain Dyce hap-

pened to be in town that night, to discuss the Tilden
Campaign with the War Governor. Governor Curtin,
who was naturally too busy with his Russian guests
to talk politics, smilingly told the Clinton County
leader that he could do him a great favor if the next
morning at eight o'clock he would have five or six
two-horse surreys in front of the Curtin home. Dyce
took the hint and spent the entire night among the
local liverymen and horse jockeys getting together
the equipment. Next morning, which dawned de-
lightfully clear, at seven-thirty found six dignified-
looking two-horse surreys, each driven by a grinning
Negro, lined up on the hilly street before the War Governor's domicile. As the party emerged from the
house the Governor addressed them, saying, "Gentlemen, we go this morning to the greatest natural wonder in Pennsylvania." The Russian dignitaries, who
were great horse lovers, spent fully fifteen minutes
inspecting the livery nags, a goodly lot of trotting-bred
type, which they declared were on the same general
lines of their own Orloffs. As he got in his carriage,
Field Marshal von Fersen, who owned a vast stock
farm on the Volga, shook his head sadly, saying,
"What a pity you Americans don't keep your horses
entire." Frequently on the drive the distinguished
tourists uttered exclamations of delight at the grand
scenery and prosperous looking farms, but they were
kept laughing most of the way at the jokes and humor-
ous anecdotes told them by Governor Curtin and Captain Dyce, both of whom had inherited inimitable wit

from their Celtic ancestors. Arriving at the Penn's
Cave Farm, the party was cordially received by Proprietor George Long and wife. Mr. Long was one
of Governor Curtin's political admirers, consequently
he whispered to his spouse to prepare the best dinner
she knew how. While it was gotten ready, the party,
led by the proprietor, was taken through the cavern in
a huge flatboat. The emotional Russians kept shouting with approbation, while Count Hickoff, who was a
fine singer, woke the echoes with the Russian National
Anthem. The visit to the dry cave was particularly
edifying to all concerned. Count Hickoff collected a
pocketful of bones and shells, while the hospitable proprietor Long broke off for him several of the choicest
stalactites. "You say that this is Pennsylvania's greatest natural wonder?" said Baron de Toplitz-Herber-
stain, as the party emerged into the warm sunlight;
"but I say that there is nothing finer in Russia, or perhaps in the world." At these words Governor Curtin
smiled, as he was an early believer in the theory of
"seeing America first," and dearly loved his native
Central Pennsylvania. The Pennsylvania country
dinner served by Mrs. Long and her hand-maidens
was fully up to the traditions of such a repast. It is
stated that nine kinds of pie were on the table at one
time. And each Russian sampled them all. Before
going to the cave Governor Curtin had explained the
"caste" system of Russia to the Longs, consequently
only the three grandees, their secretaries, the Governor
and Captain Dyce sat down to the "first table." The

Russians conversed with the Longs in High German, being replied to in Pennsylvania Dutch. The rest of the party, including the drivers and the Long family, were at the second table, and there was a-plenty for all. After the dinner, which was equal to any Russian wedding feast, all averred, the party was driven back to Bellefonte. After spending another night under Governor Curtin's hospitable roof, the happy Russians departed for Altoona and Pittsburg, loaded with letters of introduction from their host, to the car builders and steel magnates whose works they wished to inspect. But in all their travels, interesting as they doubtless were, they hardly enjoyed themselves more than their trip to "Pennsylvania's greatest natural wonder," Penn's Cave.

VIII.   THE FOUNTAIN OF YOUTH.

OLD CHIEF WISAMEK, of the Kittochtinny Indians, had lost his spouse. He was close to sixty years of age, which was old for a red man, especially one who had led the hard life of a warrior, exposed to all kinds of weather, fasts and forced marches. Though he felt terribly lonely and depressed in his state of widowerhood, the thought of discarding the fidelity of the eagle, which if bereaved never takes a second mate, and was the noble bird he worshiped was repugnant to him until he happened to see the fair and buxom maid Annapalpeteu. He was rheumatic, walking with difficulty; he tired easily, was fretful, all sure signs of increasing age; but what upset him most was the sight of his reflection in his favorite pool, a haggard, weezened, wrinkled face, with a nose like the beak of an eagle, and eyes as colorless as clay. When he opened his mouth, the reflected image seemed to be mostly toothless, the lips were blue and thin. He had noticed that he did not need to shave his skull any more to give prominence to his warrior's top-knot; the proud tuft itself was growing sparse and weak; to keep it erect he was now compelled to braid with it hair from the buffalo's tail. Brave warrior that he was, he hated to pay his court to the lovely Annapalpeteu when on all sides he saw stalwart six-foot youths, masses of sinews and muscle, clear-eyed, firm-lipped, always ambitious and high-spirited. But one afternoon he saw his copper-colored love sitting by the side of the Bohundy Creek, beating maize in a wooden trough. Her entire costume consisted of a tight petticoat of blue cloth, hardly reaching to the knees, and without any ruffles. Her cheeks and forehead were neatly daubed with red. She seemed very well content with her coadjutor, a bright young fellow, who, except for two wild cat hides appropriately distributed, was quite as naked as the ingenuous beauty. That Annapelpeteu had a cavalier was now certain, and immediately it rekindled what flames remained in his jaded body; he must have her at any cost. Down by the Conadogwinet, across the South Mountains, lived Mbison, a wise man. Old Wisamek would go there and consult him, perhaps obtain from him some potion to permanently restore at least a few of the fires of his lost youth. Though his will-power had been appreciably slackening of late years, he acted with alacrity on the idea of visiting the soothsayer. Before sundown he was on his way to the south, accompanied by several faithful henchmen. Carrying a long ironwood staff, he moved on with unwonted agility; it was very dark, and the path difficult to follow, when he finally consented to bivouac for the night. The next morning found him so stiff that he could hardly clamber to his feet. His henchmen assisted him, though they begged him to rest for a day. But his will forced him on; he wanted to be virile and win the beautiful Annapalpateu. The journey, which consumed a week, cost the aged Strephon a

world of effort. But as he had been indefatigable in
his youth, he was determined to reach the wise man's
headquarters walking like a warrior, and not carried
there on a litter like an old woman. Bravely he forged
ahead, his aching joints paining miserably, until at
length he came in sight of his Promised Land. The
soothsayer, who had been apprsed of his coming by a
dream, was in front of his substantial lodge-house
to greet him. Seldom had he received a more distinguished client than Wisamek, so he welcomed him
with marked courtesy and deference. After the first
formalities, the old chief, who had restrained himself with difficulty, asked how he could be restored to
a youthful condition so that he could rightfully marry
a beautiful maiden of eighteen summers. The wise
man, who had encountered similar supplicants in the
past, informed him that the task was a comparatively
easy one. It would involve, however, another journey
across mountains. Wisamek shouted for joy when
he heard these words and impatiently demanded
where he would have to go to be restored to youth.
"Across many high mountain ranges, across many
broad valleys, across many swift streams, through a
country covered with dark forests and filled with wild
beasts, to the north-west of here is a wonderful cavern.
In it rises a deep stream, of greenish color, clear as
crystal, the fountain of youth. At its heading you will
find a very old man, Gamunk, who knows the formula.
Give him this talisman, and he will allow you to bathe
in the marvelous waters, and be young again." With

THE JEWELED CHAMBER.

the final words he handed Wisamek a red bear's
tooth, on which was cleverly carved the form of an
athletic youth. The old chief's hands trembled so
much that he almost dropped the precious fetich. But
he soon recovered his self-control and thanked the
wise man. Then he ordered his henchmen to give the
soothsayer gifts, which they did, loading him with
beads, pottery, wampum and rare furs. Despite the
invitation to remain until he was completely rested,
Wisamek determined to depart at once for the fountain of youth. He was so stimulated by his high hope
that he climbed the steep ridges, crossed the turbulent
streams, and put up with the other inconveniences of
the long march much better than might have been the
case. During the entire journey he sang Indian love
songs, strains which had not passed his lips in thirty
years. His followers, gossiping among themselves,
declared that he looked better already. Perhaps he
would not have to bathe in the fountain after all. He
might resume his youth, because he willed it so.
Indians were strong believers in the power of mind
over matter. When he reached the vicinity of the cave
he was fortunate enough to meet the aged Indian who
was its guardian. Though his hair was snow white,
and he said he was so old that he had lost count of the
years, Gamunk's carriage was erect, his complexion
smooth, his eyes clear and kindly. He walked along
with a swinging stride, very different from Wisamek's
mental picture of him. The would-be bridegroom,
who handed him the talisman, was quick to impart

his mission to his new-found friend. "It is true,"
he replied; "after a day and a night's immersion in
the cave's water you will emerge with all the appearance of youth. There is absolutely no doubt of it.
Thousands have been here before." With these re-
assuring words Wisamek again leaped for joy, gyrat-
ing like a young brave at a cantico. The party, accompanied by the old guardian, quickly arrived at
the cave's main opening, where beneath them lay
stretched the calm, mirror-like expanse of greenish
water. "Can I begin the bath now?" asked the chief,
impatiently. "I am anxious to throw off the odious
appearance of age." "Immediately," replied the old
watchman, who took him by the hand, leading him to
the ledge where it was highest above the water.
"Jump off here," he said quietly. Wisamek, who had
been a great swimmer in his youth and was absolutely
fearless of the water, replied that he would do so.
"But remember you must remain in the water without
food until this hour tomorrow," said the guardian.
As he leaped into the watery depths the chief shouted
he would remain twice as long if he could be young
again. Wisamek was true to his instructions; there
was too much at stake; he dared not falter. The next
morning his henchmen were at the cave's mouth to
greet his reappearance. They were startled to see,
climbing up the ledge with alacrity a tall and handsome man, as young looking as themselves. There
was a smile on the full red lips, a twinkle in the clear
eye of the re-made warrior as he stood among them,

physically a prince among men. The homeward journey was made with rapidity. Wisamek traveled so
fast that he played out his henchmen who were half
his age. Annapalpeteu, who was seated in front of
her parents' cabin, weaving a garment, noticed a
youth of great physical beauty approaching, at the
head of Chief Wisamek's clansmen. She wondered
who he could be, as he wore Wisamek's headdress
of feathers of the sea eagle. When he drew near he
saluted her, and, not giving her time to answer, joy-
fully shouted, "Don't you recognize me? I am your
good frind Wisamek, come back to win your love,
after a refreshing journey through the distant forests." Annapalpeteu, who was a sensible enough girl
to have admired the great warrior for his prowess,
even though she had never thought of him seriously
as a lover, was now instantly smitten by his engaging
appearance. The henchmen withdrew, leaving the
couple together. They made marked progress with
their romance; words of love were mentioned before
they parted. It was not long before the betrothal was
announced, followed shortly by the wedding festival.
At the nuptials the bridegroom's appearance was the
marvel of all present. It was hinted that he had been
somewhere and renewed his youth, but as the henchmen were sworn to secrecy, how it had been done was
not revealed. The young bride seemed radiantly
happy. She had every reason to be; the other Indian
maids whispered from lip to lip, was she not marrying
the greatest warrior and hunter of his generation, the

handsomest man in a hundred tribes? Secretly envied by all of her age, possessing her stalwart prize,
the fair bride started on her honeymoon, showered
with acorns and good wishes. So far as is known the
wedding trip passed off blissfully. There were smiles
on the bright faces of both bride and groom when
they returned to their spacious new lodge-house, which
the tribe had erected for them in their absence, by the
banks of the rippling Bohundy. But the course of life
did not run smoothly for the pair. Though outwardly
Wisamek was the handsomest and most youthful
looking of men, he was still an old man at heart.
Annapalpeteu was as pleasure-loving as she was beautiful. She wanted to dance and sing and mingle with
youthful company. She wanted her good time in life;
her joy of living was at its height, her sense of enjoy-
ment at its zenith. On the other hand, Wisamek
hated all forms of gaieties or youthful amusements.
He wanted to sit about the lodge-house in the sun,
telling of his warlike triumphs of other days; he
wanted to sleep much, he hated noise and excitement.
Annapalpeteu, dutiful wife that she was, tried to
please him, but in due course of time both husband
and wife realized that romance was dying, that they
were difting apart. Wisamek was even more aware
of it than his wife. It worried him greatly, his
dreams were of an unhappy nature. He pictured the
end of it all, with his wife, Annapalpeteu, in love with
some one else of her own age, some one whose heart
was young. He had spells of moodiness and irrita-

bility, as well as several serious quarrels with his wife, whom he accused of caring less for him than formerly. The relations became so strained that life in the commodious lodge-house was unbearable. At length it occurred to Wisamek that he might again visit the fountain of youth, this time to revive his soul. Perhaps he had not remained in the water long enough to touch the spirit within. He informed his spouse that he was going on a long journey, on invitation of the chief of a distant tribe, and that she must accompany him. He was insanely jealous of her now; he could not bear her out of his sight. He imagined she had a young lover hiding back of every tree, though she was honor personified. The trip was made pleasantly enough, as the husband was in better spirits than usual. He thought he saw the surcease of his troubles ahead of him! When he reached the Beaver Dam Meadows, near the site of the present town of Spring Mills, beautiful level flats which in those days were a favorite camping ground for the red men, he requested the beautiful Annapalpeteu to remain there for a few days, that he was going into a hostile country, he would not jeopardize her safety. He was going on an important mission that would make her love him more than ever when he returned. In reality no unfriendly Indians were about, but in order to give a look of truth to his story, he left her in charge of a strong bodyguard. Wisamek's conduct of late had been so peculiar that his wife was not sorry to see her lord and master go away. Handsome though he was, a spiritual barrier had arisen between them which grew more insurmountable with each succeeding day. Yet, on this occasion, when he was out of her sight, she felt apprehensive about him. She had a strange presentment that she would never see him again. Wisamek was filled with hopes; his spirits had never been higher, as he strode along, followed by his henchmen. When he reached the top of the path which led to the mouth of the cave he met old Gamunk, the guardian. The aged red man expressed surpise at seeing him again. "I have come for a very peculiar reason," he said. "The bath which I took last year outwardly made me young, but only outwardly. Within I am as withered and joyless as a centenarian. I want to bathe once more, to try to revive the old light in my soul." Gamunk shook his head. "You may succeed; I hope you will. I never heard of any one daring to take a second bath in these waters. The tradition of the hereditary guardians, of whom I am the hundredth in direct succession, has it that it would be fatal to take a second immersion, espcially to remain in the water for twenty-four hours." Then he asked Wisamek for the talisman which was the right to bathe. Wicamek drew himself up proudly, and with a gesture of his hand, indicating disdain, said he had no talisman, that he would bathe anyhow. He advanced to the brink and plunged in. Until the same hour the next day he floated and paddled about the greenisn depths, filled with expectancy. For some reason it seemed longer this time than on the previous visit.

At last, by the light which filtered down through the treetops at the cave's mouth, he knew that the hour had come for him to emerge —emerge as Chief Wisamek—young in heart as in body. Proudly he grasped the rocky ledge, and swung himself out on dry land. He arose to his feet. His head seemed very light and giddy. He fancied he saw visions of his old conquests, old loves. There was the sound of music in the air. Was it martial music, played to welcome the conqueror, or the wind surging through the feathery tops of the maple and linden trees at the mouth of the cave? He started to climb the steep path. He seemed to be treading on air. Was it the buoyant steps of youth come again? He seemed to float rather than walk. The sunlight blinded his eyes. Suddenly he had a flash of normal consciousness. He dropped to the ground with a thud like an old pine falling. Then all was blackness, silence. Jaybirds complaining in the trees alone broke the stillness. His bodyguards, who were waiting for him at old Gamunk's lodgehouse, close· to where the hotel now stands, became impatient at his non-appearance, as the hour was past. Accompanied by the venerable watchman, they started down the path. To their horror they saw the dead body of a hideous, wrinkled old man, all skin and bones, lying stretched out across it, a few steps from the entrance to the cave. When they approached closely they noticed several familiar tattoo marks which identified the body as that of their late master, Wisamek. Frightened lest they would be accused of his murder, and shocked by his altered appearance, the bodyguards turned and took to their heels. They disappeared in the trackless forests to the north and were never seen again. Old Gamunk, out of pity for the vainglorious chieftain, buried the remains by the path near where he fell. As for poor Annapalpeteu, the beautiful, she waited patiently for many days by the Beaver Dams, but her waiting was in vain. At length, concluding that he had been slain in battle in some valorous encounter, she started for her old home on the Bohundy. It is related that in due course of time she married a warrior of her own age, living happily ever afterwards. In him she found the loving response, the congeniality of pleasures which had been denied the dried, feeble soul of Wisamek, who bathed once too often in the fountain of youth.

SCENERY FROM PENN'S CAVE PROPERTY

IX.   RIDING HIS PONY.

WHEN Rev. James Martin visited Penn's Cave, in the Spring of 1795, it was related that he found a small group of Indians encamped there. That evening, around the campfire, one of the redskins related a legend of one of the curiosities of the watery cave, the flamboyant "Indian Riding Pony" mural-piece which decorates one of the walls. Spirited as a Remington, it bursts upon the view, creates a lasting impression, then vanishes as the power skiff, the "Nita-nee," draws nearer. According to the old Indians, there lived not far from where the Karoondinha emerges from the cavern a body of savages who made this delightful lowland their permanent abode. While most of their cabins were huddled near together on the upper reaches of the stream, there were straggling huts clear to the Beaver Dams. The finding of arrow points, beads and pottery along the creek amply attests to this. Among the clan was a maiden named Quetajaku, not good to look upon, but in no way ugly or deformed. In her youth she was light-hearted and sociable, with a gentle disposition. Yet for some reason she was not favored by the young bucks. All her contemporaries found lovers and husbands, but poor Quetajaku was left severely alone. She knew that she was not beautiful, though she was of good size; she was equally certain that she was not a physical monster. She could not understand why she could find no lover. why she was singled out to be a "chauch-schisis," or old maid. It hurt her pride as a young girl, it broke her heart completely when she was older. Gradually she withdrew from the society of her tribal friends, building herself a lodge-house on the hill, in what is now the cave orchard. There she led a very introspective life, grieving over the love that might have been. To console herself she imagined that some day a handsome warrior would appear, seek her out, load her with gifts, overwhelm her with love, and carry her away to some distant region in triumph. He would be handsomer and braver than any youth in the whole country of the Karoondinha. She would be the most envied of women when he came. This poor little fancy saved her from going stark mad, it remedied the horror of her lonely lot. Every time the night wind stirred the rude cloth which hung before the door of her cabin, she would picture it was the chivalrous stranger come to claim her. When it was cold she drew the folds of her buffalo robe tighter about her as if it was his arms. As time went on she grew happy in her secret lover, whom no other woman's flame could equal, whom no one could steal away. She was ever imagining him saying to her that her looks exactly suited him, that she was his ideal. But like the seeker after Eldorado, years passed, and Quetajaku did not come nearer to her spirit lover. But her soul kept up the conceit; every night when she curled herself up to sleep he was the vastness of the night. On one occasion an Indian artist named Niganit, an undersized old wanderer, appeared at the lonely woman's home. For a living he decorated pottery, shells and bones, sometimes even painted war pictures on rocks Quetajaku was so kind to him that he built himself a lean-to on the slope of the hill, intending to spend the winter. On the long winter evenings the old woman confided to the wanderer the story of her unhappy life, of her inward consolation. She said that she had longed to meet an artist who could carry out a certain part of her dream which had a right to come true. When she died she had arranged to be buried in a fissure of rocks which ran horizontally into one of the walls of the "watery" cave. On the opposite wall she would like painted in the most brilliant colors a portrait of a handsome young warrior, with arms outstretched, coming towards her. Niganit said that he understood what she meant exactly, but suggested that the youth be mounted on a pony, a beast which was coming into use as a mount for warriors, of which he had lately seen a number in his travels on the Virginia coast. This idea was pleasing to Quetajaku, who authorized the stranger to begin work at once. She had saved up a little property of various kinds; she promised to be-stow all of this on Niganit, except what would be necessary to bury her, if the picture proved satisfactory. The artist rigged up a dog-raft with a scaffold on it, and this he poled into the place where the fissure was located, the woman accompanying him the first time, so there would be no mistake. All winter long by torchlight he labored away. He used only one color, an intensive brick-red made from mixing sumac, a kind of seed, a small root and the bark of a tree, as being more permanent than that made from ochers and other ores or stained earth. Marvelous and vital was the result of this early impressionist; the painting had all the action of life. The superb youth in war dress, with arms outstretched, on the agile war pony, was rushing towards the foreground, almost in the act of leaping from the rocky panel into life, across the waters of the cave to the arms of his beloved. It would make old Quetajaku happy to see it, she who had never known love or beauty. The youth in the mural typified what Niganit would have been himself were he the chosen, and what the old squaw would have possessed had nature favored her. It was the ideal for two disappointed souls. Breathlessly the old artist ferried Quetajaku to the scene of his endeavors. When they reached the proper spot he held aloft his quavering torch. Quetajaku, in order to see more clearly, held her two hands above her eyes. She gave a little cry of exclamation, then turned and looked at Niganit intently. Then she dropped her eyes, beginning to cry to herself. The artist looked at her fine face, down which the tears were streaming, and asked her the cause of her grief—was the picture such a terrible disappointment? The woman drew herself together, replying that it was grander than she had anticipated, but the face was Niganit's, and, strangely enough, was the face she had dreamed of all her life. "But I am

GRAVE OF REV. JAMES MARTIN (near Penn Hall, Centre County.)
One of the first white men to enter Penn's Cave.

not the heroic youth you pictured," said the artist, sadly. "I am sixty years old, stoop-shouldered, and one leg is shorter than the other." "But that is how you would look on your war pony; it is your face, shoulders and arms. You are the picture that I always hoped would come true." Niganit looked at the Indian woman. She was not *hideous;* there was even a dignity to her large, plain features, her great, gaunt form. I"have never received praise such as yours. I always vowed I would love the woman who really understood me and my art. I am yours. Let us think no more of funeral decorations, but go to the east, to the land of the war ponies, and ride to endless joy together." Quetajaku, overcome by the majesty of his words, leaned against his massive shoulder. In that way he poled his dog-raft against the current to the entrance of the cave. There was a glory in the reflection from the setting sun over against the east; night would not set in for an hour or two. And towards the darkening east that night two happy travelers could be seen wending their way.

X.   NITA-NEE.

The Indian Maiden for Whom Nittany Mountain is Named.

(Reprinted from "Juniata Memories," Philadelphia, 1916.)

(Copyrighted by J. J. McVey, Publisher.)

ONE of the last Indians to wander through the Juniata Valley, either to revive old memories or merely to hunt and trap, his controlling motive is not certain, was old Jake Faddy. As he was supposed to belong to the Seneca tribe, and spent most of his time on the Coudersport Pike on the border line between Clinton and Potter Counties, it is to be surmised that he never lived permanently on the Juniata, but had hunted there or participated in the bloody wars in the days of his youth. He continued his visits until he reached a very advanced age. Of a younger generation than Shaney John, he was nevertheless well acquainted with that unique old redman, and always spent a couple of weeks with him at his cabin on Saddler's Run.

Old Jake, partly to earn his board and partly to show his superior knowledge, was a gifted story teller. He liked to obtain the chance to spend the night at farmhouses where there were aged people, and his smattering of history would be fully utilized to put the older folks in good humor.

For while the hard-working younger generations
fancied that *history* was a waste of time, the old people
loved it, and fought against the cruel way in which all
local tradition and legend was being snuffed out.   If
it had not been for a few people carrying it over the
past generation, all of it would now be lost in the whirl-
pool of a commercial, materialistic age. And to those
few, unknown to fame, and of obscure life and residence, is due the credit of saving for us the wealth of
folklore that the noble mountains, the dark forests, the
wars and the Indians, instilled in the minds of the first
settlers.   And there is no old man or woman living in
the wilderness who is without a story that is ready to
be imparted, and worthy of preservation.   But the
question remains, how can these old people all be
reached before they pass away?   It would take an
army of collectors, working simultaneously, as the
Grim Reaper is hard at work removing these human
landmarks with their untold stories.

Out near the heading of Beaver Dam Run, at the
foot of Jack's Mountain, stands a very solid-looking
stone farmhouse, a relic of pioneer days.   Its earliest
inhabitants had run counter to the Indians of the
neighborhood for the possession of the beavers whose
dens and "cabins" were its most noticeable feature
clear to the mouth of the stream, and later for the
otters who defied the white annihilators a quarter of
a century longer.   Beaver trapping had made the
stream a favorite rendezvous for the red men, and
their campgrounds at the springs near the headwaters

were pointed out until a comparatively recent date.

But one by one the aborigines dropped away, until Jake Faddy alone upheld the traditions of the race. There were no beavers to quarrel over in his day, consequently his visits were on a more friendly basis. The old North of Ireland family who occupied the stone farmhouse was closely linked with the history of the Juniata Valley, and they felt the thrill of the vivid past whenever the old Indian appeared at the kitchen door. As he was ever ready to work and, what was better, a very useful man at gardening and flowers, he was always given his meals and lodging for as long as he cared to remain. But that was not very long, as his restless nature was ever-goading him on, and he had "many other friends to see," putting it in his own language. He seemed proud to have it known that he was popular with a good class of white people, and his ruling passion may have been to cultivate these associations. On several occasions he brought some of his sons with him, but they did not seem anxious to live up to their father's standards. And after the old man had passed away none of this younger generation ever came to the Juniata Valley.

The past seemed like the present to Jake Faddy, he was so familiar with it. To him it was as if it happened yesterday, the vast formations and changes and epochs. And the Indian race, especially the eastern Indians, seemed to have played the most important part in those titanic days. It seemed so recent and so real to the old redman that his stories were always interest-

WHERE REV. MARTIN WROTE HIS SERMONS
(Near Penn Hall.)

ing. The children also were fond of hearing him talk;
he had a way of never becoming tiresome. Every
young person who heard him remembered what he
said. There would have been no break in the "apos-
tolic succession" of Pennsylvania legendary lore if all
had been seated at Jake Faddy's knee.

Of all his stories, by odds his favorite one, dealt
with the Indian maiden, Nita-nee, for whom the fruit-
ful Nittany Valley and the towering Nittany Mountain are named. This Indian girl was born on the
banks of the lovely Juniata, not far from the present
town of Newton Hamilton, the daughter of a powerful
chief. It was in the early days of the world, when
the physical aspect of Nature could be changed over
night by a fiat from the Gitchie-Manitto or Great
Spirit. It was therefore in the age of great and wonderful things, before a rigid world produced beings
whose lives followed grooves as tight and permanent
as the gullies and ridges.

During the early life of Nita-nee a great war was
waged for the possession of the Juniata Valley. The
aggressors were Indians from the South, who longed
for the scope and fertility of this earthly Paradise.
Though Nita-nee's father and his brave cohorts de-
fended their beloved land to the last extremity, they
were driven northward into the Seven Mountains and
beyond. Though they found themselves in beautiful
valleys, filled with bubbling springs and teeming with
game, they missed the Blue Juniata, and were never
wholly content. The father of Nita-nee, who was

named Chun-Eh-Hoe, felt so humiliated that he only went about after night in his new home. He took up his residence on a broad plain, not far from where State College now stands, and should be the Indian patron of that growing institution, instead of Chief Bald Eagle, who never lived near there and whose good deeds are far outweighed by his crimes.

Chun-Eh-Hoe was an Indian of exact conscience. He did his best in the cruel war, but the southern Indians must have had more sagacious leaders or a better *esprit de corps*. At any rate they conquered. Chun-Eh-Hoe was not an old man at the time of his defeat, but it is related that his raven black locks turned white over night. He was broken in spirit after his downfall and only lived a few years in his new home. His widow, as well as his daughter, Nita-nee, and many other children, were left to mourn him. As Nita-nee was the oldest, she assumed a vicereineship over the tribe until her young brother, Wo-Wi-Na-Pie, should be old enough to rule the councils and go on the warpath.

The defeat on the Juniata, the exile to the northern valleys and the premature death of Chun-Eh-Hoe were to be avenged. Active days were ahead of the tribesmen. Meanwhile if the southern Indians crossed the mountains to still further covet their lands and liberties, who should lead them to battle but Nita-nee. But the Indian vicereine was of a peace-loving disposition. She hoped that the time would never come when she would have to preside over scenes of carnage and slaughter. She wanted to see her late father's tribe become the most cultured and prosperous in the Indian world, and in that way be revenged on their warlike foes: "Peace hath its victories."

But she was not to be destined to lead a peaceful nation through years of upward growth. In the Juniata Valley the southern Indians had become over-populated; they sought broader territories, like the Germans of today. They had driven the present occupants of the northern valleys out of the Juniata country, they wanted to again drive them further north.

Nita-nee did not want war, but the time came when she could not prevent it. The southern Indians sought to provoke a conflict by making settlements in the Bare Meadows, and in some fertile patches on Tussey Knob and Bald Top, all of which were countenanced in silence. But when they murdered some peaceable farmers and took possession of plantations at the foot of the mountains in the valley of the Karoondinha, then the mildness of Nita-nee's cohorts came to an end. Meanwhile her mother and brother had died, Nita-nee had been elected queen.

Every man and boy volunteered to fight; a huge army was recruited over night. They swept down to the settlements of the southern Indians, butchering every one of them. They pressed onward to the Bare Meadows, and to the slopes of Bald Top and Tussey Knob. There they gave up the population to fire and sword. Crossing the Seven Mountains, they formed a powerful cordon all along the southerly slope of the

Long Mountain. Building block houses and stone fortifications—some of the stonework can be seen to this day—they could not be easily dislodged.

The southern Indians, noticing the flames of the burning plantations, and hearing from the one or two survivors of the completeness of the rout, were slow to start an offensive movement. But as Nita-nee's forces showed no signs of advancing beyond the foot of Long Mountain, they mistook this hesitancy for cowardice, and sent an attacking army. It was completely defeated in the gorge of Laurel Run, above Milroy, and on the slopes of Sample Knob, the right of the northern Indians to the Karoondinha and the adjacent valleys was signed, sealed and delivered in blood. The southern Indians were in turn driven out by other tribes; in fact, every half century or so a different race ruled over the Juniata Valley. But in all those years none of the Juniata rulers sought to question the rights of the northern Indians until 1635, when the Lenni-Lenape invaded the country of the Susquehannocks and were decisively beaten on the plains near Rock Spring, in Spruce Creek Valley, at the Battle of the Indian Steps. (This battle has been described in stirring verse by Central Pennsylvania's bard, John H. Chatham, "The Indian Steps," Altoona, 1913.)

As Nita-nee wanted no territorial accessions, she left the garrisons at her southerly forts intact, and retired her main army to its home valleys, where it was disbanded as quickly as it came together. All were glad to be back to peaceful avocations, none of them

craved glory in war. And there were no honors given
out, no great generals created. All served as private
soldiers under the direct supervision of their queen. It
was the theory of this Joan of Arc that by eliminating
titles and important posts there would be no military
class created, no ulterior motive assisted except *pa-
triotism*. The soldiers serving anonymously, and for
their country's need alone, would be ready to end their
military duties as soon as their patriotic task was done.

Nita-nee regarded soldiering as a stern necessity,
not as an excuse for pleasure or pillage, or personal
advancement. Under her there was no nobility, all
were on a common level of dignified citizenship. Every
Indian in her realm had a task, not one that he was
born to follow, but the one which appealed to him
mostly, and therefore the task at which he was most
successful. Women also had their work, apart from
domestic life in this ideal democracy of ancient days.
Suffrage was universal to both sexes over twenty years
of age, but as there were no official positions, no public
trusts, a political class could not come into existence,
and the queen, as long as she was canning and able,
had the unanimous support of her people. She was
given a great ovation as she modestly walked along
the fighting line after the winning battle of Laurel
Run. It made her feel not that she was great, but
that the democracy of her father and her ancestors was
a living force. In those days of pure democracy the
rulers walked: the litters and palanquins were a later
development.

After the conflict the gentle Nita-nee, at the head of the soon to be disbanded army, marched across the Seven Brothers, and westerly toward her permanent encampment, where State College now stands. As her only trophy she carried a bundle of spears, which her brave henchmen had wrenched from the hands of the southern Indians as they charged the forts along Long Mountain. These were not to deck her own lodge house, nor for vain display, but were to be placed on the grave of her father, the lamented Chun-Eh-Hoe, who had been avenged. In her heart she had hoped for victory, almost as much for his sake as for the comfort of her people. She knew how he had grieved himself to death when he was outgeneraled in the previous war.

In theose dimly remote days there was no range of mountains where the Nittany chain now raise their noble summits to the sky. All was a plain, a prairie, clear north to the Bald Eagles, which only recently had come into existence. The tradition was that far older than all the other hills were the Seven Mountains. And geological speculation seems to bear this out. At all seasons of the year cruel and chilling winds blew out of the north, hindering the work of agriculture on the broad plains ruled over by Nita-nee. Only the strong and the brave could cope with these killing blasts, so intense and so different from the calming zephyrs of the Juniata. The seasons for this cause were several weeks shorter than across the Seven Mountains; that is, there was a later spring and an earlier fall. But though the work was harder, the soil being equally rich and broader area, the crops averaged fully as large as those further south. So, taken altogether, the people of Nita-nee could not be said to be an unhappy aggregation.

As the victorious queen was marching along at the head of her troops, she was frequently almost mobbed by women and children, who rushed out from the settlements and made her all manner of gifts. As it was in the early spring, there were no floral garlands, but instead wreaths and festoons of laurel, of ground pine and ground spruce. There were gifts of precious stones and metals, of rare furs, of beautiful specimens of Indian pottery, basketry and the like. These were graciously acknowledged by Nita-nee, who turned them over to her bodyguards to be carried to her permanent abode on the "Barrens." But it was not a "barrens" in those days, but a rich agricultural region, carefully irrigated from the north, and yielding the most bountiful crops of Indian corn. It was only when abandoned by the frugal redmen and grown up with forest which burned over repeatedly through the carelessness of the white settlers that it acquired that disagreeable name. In those days it was known as the "Hills of Plenty."

As Nita-nee neared the scenes of her happy days she was stopped in the middle of the path by an aged Indian couple. Leaning on staffs in order to present a dignified appearance, it was easily seen that age had bent them nearly double. Their weazened, weather-beaten old faces were pitiful to behold. Toothless, and barely able to speak above a whisper, they addressed the gracious queen.

"We are very old," they began, "the winters of more than a century have passed over our heads. Our sons and our grandsons were killed fighting bravely under your immortal sire, Chun-Eh-Hoe. We have had to struggle on by ourselves as best we could ever since. We are about to set out a crop of corn, which we need badly. For the past three years the north wind has destroyed our crop every time it appeared; the seeds which we plan to put in the earth this year are the last we've got. Really we should have kept them for food, but we hoped that the future would treat us more generously. We would like a wind-break built along the northern side of our corn patch; we are too feeble to go to the forests and cut and carry the poles. Will not our most kindly queen have some one assist us:

Nita-nee smiled on the aged couple, then she looked at her army of able-bodied warriors.

Turning to them she said, "Soldiers, will a hundred of you go to the nearest royal forest, which is in the center of this plain, and cut enough cedar poles with brush on them to build a wind-break for these good people?"

Instantly a roar arose, a perfect babel of voices; it was every soldier trying to volunteer for this philanthropic task.

When quiet was restored, a warrior stepped out

PENN'S CAVE HOTEL.

from the lines saying, "Queen, we are very happy to do this, we who have lived in this valley know full well how all suffer from the uncheckable north winds."

The queen escorted the old couple back to their humble cottage, and sat with them until her stalwart braves returned with the green-tipped poles. It looked like another Birnam Wood in process of locomotion. The work was so quickly and so carefully done that it seemed almost like a miracle to the wretched old Indians. They fell on their knees, kissing the hem of their queen's garment and thanking her for her beneficence. She could hardly leave them, so profuse were they in their gratitude. In all but a few hours were consumed in granting what to her was a simple favor, and she was safe and sound within her royal lodge house by dark. Before she left she had promised to return when the corn crop was ripe and partake of a corn roast with the venerable couple. The old people hardly dared hope she would come, but those about her knew that her word was as good as her bond. That night bonfires were lighted to celebrate her return, and there was much Indian music and revelry.

Nita-nee was compelled to address the frenzied mob, and in her speech she told them that while they had won a victory, she hoped it would be the last while she lived; she hated war, but would give her life rather than have her people invaded. All she asked in this world was peace with honor. That expressed the sentiment of her people exactly, and they literally went mad with loyalty and enthusiasm for the balance of the night. Naturally with such an uproar there was no sleep for Nita-nee.

As she lay awake on her couch she thought that far sweeter than victory or earthly fame was the helping of others, the smoothing of rough pathways for the weak or oppressed. She resolved more than ever to dedicate her life to the benefiting of her subjects. No love affair had come into her life, she would use her great love-nature to put brightness into unhappy souls about her. And she got up the next morning much more refreshed than she could have after a night of sleep surcharged with dreams of victory and glory.

As the summer progressed, and the corn crop in the valleys became ripe, the queen sent an orderly to notify the aged couple that she would come to their home alone the next evening for the promised corn roast. It was a wonderful, calm, cloudless night, with the full moon shedding its effulgent smile over the plain. Unaccompanied, except by her orderly, Nita-nee walked to the modest cabin of the aged couple, a distance of about five miles, for the cottage stood not far from the present village of Linden Hall. Evidently the windbreak had been a success, for, bathed in moonlight, the tasseled heads of the cornstalks appeared above the tops of the cedar hedge. Smoke was issuing from the open hearth back of the hut, which showed that the roast was being prepared. The aged couple were delighted to see her, and the evening passed by, bringing innocent and supreme happiness to all. And thus in broad unselfishness and generosity of thought and

deed the great queen's life was spent, making her path-
way through her realm radiant with sunshine.

And when she came to die, after a full century of
life, she requested that her body be laid to rest in the
royal forest, in the center of the valley whose people
she loved and served so well. Her funeral cortege,
which included every person in the plains and valleys,
a vast assemblage, shook with a common grief. It
would be hard to find a successor like her, a pure
soul so deeply animated with true godliness.

And it came to pass that on the night when she
was buried beneath a modest mound covered with
cedar boughs, and the vast funeral party had dispersed,
a terriffic storm arose, greater than even the oldest person could remember. The blackness of the night was
intense, the roar and rumbling heard made every being fear that the end of the world had come. It was
a night of intense terror, of horror. But at dawn, the
tempest abated, only a gentle breeze remained, a
golden sunlight overspread the scene, and great was
the wonder thereof! In the center of the vast plain
where Nita-nee had been laid away stood a mound-
like mountain, a towering, sylvan giant covered with
dense groves of cedar and pine. And as it stood there,
eternal, it tempered and broke the breezes from the
north, promising a new prosperity, a greater tran-
quility, to the peaceful dwellers in the newly-created
vale that has since been called the Valley of the
Karoondinha.

A miracle, a sign of approval from the Great Spirit, had happened during the night to forever keep alive the memory of Nita-nee, who had tempered the winds from the cornpatch of the aged, helpless couple years before. And the dwellers in the valleys adjacent to the now protected Valley of the Karoondinha awoke to a greater pride in themselvs, a high ideal must be observed, since they were the special objects of celestial notice.

And the name of Nita-nee was the favorite cognomen for Indian maidens, and has been borne by many of saintly and useful life ever since, and none of these namesakes were more deserving than the Nita-nee who lived centuries later near the mouth of Penn's Cave, the daughter of Chief O-Ko-Cho.

PENN'S CAVE

Located in the beautiful Penn's Valley, which is noted for its fine mountain scenery, game and fishing.

Boat ride of one-half mile underground through the Cave, showing the eroded blue limestone arch with its stalacite and stalagmite formation.

THE HOTEL IS LOCATED ON THE PENN'S CAVE FARM

RAILROAD STATION
Rising Springs, Centre County, Pa.
On the Lewisburg & Tyrone R. R.

POST OFFICE
Centre Hall

Printed in the United States
78621LV00001B/247-276